D0635763

DISCARDED

LAKE OSWEGO JR. HIGH SCHOOL
2500 S.W. COUNTRY CLUB RD.
LAKE OSWEGO, OR 97034

BUILDING
HISTORY
SERIES

THE

SPACE

SHUTTLE

Titles in the Building History Series Include:

Alcatraz
The Atom Bomb
The Eiffel Tower
The Great Wall of China
The Medieval Castle
The Medieval Cathedral
Mount Rushmore
The New York Subway System
The Palace of Versailles
The Panama Canal
The Parthenon of Ancient Greece
The Pyramids of Giza
The Roman Colosseum
Roman Roads and Aqueducts
Shakespeare's Globe
The Statue of Liberty
Stonehenge
The Titanic
The Transcontinental Railroad
The Viking Longship
The White House

BUILDING
HISTORY
SERIES

THE

SPACE

SHUTTLE

by Robert Taylor

LAKE OSWEGO JR. HIGH SCHOOL
2500 S.W. COUNTRY CLUB RD.
LAKE OSWEGO, OR 97034
636-0335

Lucent Books, Inc., San Diego, California

Library of Congress Cataloging-in-Publication Data

Taylor, Robert, 1948–
    The space shuttle / by Robert Taylor.
        p.   cm. — (Building history series)
    Includes bibliographical references and index.
    Summary: Discusses the history and development, technologi-
cal and political challenges, and future of the world's first reusable
space vehicle, including the shuttle program's effects on NASA.
    ISBN 1-56006-722-5
    1. Space shuttles—Juvenile literature. [1. Space shuttles. 2.
United States. National Aeronautics and Space Administration.]
I. Title. II. Series.
  TL795.515 .T39 2002
  629.44'1—dc21

2001003153

Copyright 2002 by Lucent Books, Inc.
10911 Technology Place, San Diego, California, 92127

No part of this book may be reproduced or used in any other form
or by any other means, electrical, mechanical, or otherwise, including, but not limited to, photocopy, recording, or any information
storage and retrieval system, without prior written permission from
the publisher.

Printed in the U.S.A.

CONTENTS

FOREWORD

Throughout history, as civilizations have evolved and prospered, each has produced unique buildings and architectural styles. Combining the need for both utility and artistic expression, a society's buildings, particularly its large-scale public structures, often reflect the individual character traits that distinguish it from other societies. In a very real sense, then, buildings express a society's values and unique characteristics in tangible form. As scholar Anita Abramovitz comments in her book *People and Spaces*, "Our ways of living and thinking—our habits, needs, fear of enemies, aspirations, materialistic concerns, and religious beliefs—have influenced the kinds of spaces that we build and that later surround and include us."

That specific types and styles of structures constitute an outward expression of the spirit of an individual people or era can be seen in the diverse ways that various societies have built palaces, fortresses, tombs, churches, government buildings, sports arenas, public works, and other such monuments. The ancient Greeks, for instance, were a supremely rational people who originated Western philosophy and science, including the atomic theory and the realization that the earth is a sphere. Their public buildings, epitomized by Athens's magnificent Parthenon temple, were equally rational, emphasizing order, harmony, reason, and above all, restraint.

By contrast, the Romans, who conquered and absorbed the Greek lands, were a highly practical people preoccupied with acquiring and wielding power over others. The Romans greatly admired and readily copied elements of Greek architecture, but modified and adapted them to their own needs. "Roman genius was called into action by the enormous practical needs of a world empire," wrote historian Edith Hamilton. "Rome met them magnificently. Buildings tremendous, indomitable, amphitheaters where eighty thousand could watch a spectacle, baths where three thousand could bathe at the same time."

In medieval Europe, God heavily influenced and motivated the people, and religion permeated all aspects of society, molding people's worldviews and guiding their everyday actions. That spiritual mindset is reflected in the most important medieval structure—the Gothic cathedral—which, in a sense, was a model of heavenly cities. As scholar Anne Fremantle so elegantly phrases it, the cathedrals were "harmonious elevations of stone and glass reaching up to heaven to seek and receive the light [of God]."

Our more secular modern age, in contrast, is driven by the realities of a global economy, advanced technology, and mass communications. Responding to the needs of international trade and the growth of cities housing millions of people, today's builders construct engineering marvels, among them towering skyscrapers of steel and glass, mammoth marine canals, and huge and elaborate rapid transit systems, all of which would have left their ancestors, even the Romans, awestruck.

In examining some of humanity's greatest edifices, Lucent Books' Building History series recognizes this close relationship between a society's historical character and its buildings. Each volume in the series begins with a historical sketch of the people who erected the edifice, exploring their major achievements as well as the beliefs, customs, and societal needs that dictated the variety, functions, and styles of their buildings. A detailed explanation of how the selected structure was conceived, designed, and built, to the extent that this information is known, makes up the majority of the volume.

Each volume in the Lucent Building History series also includes several special features that are useful tools for additional research. A chronology of important dates gives students an overview, at a glance, of the evolution and use of the structure described. Sidebars create a broader context by adding further details on some of the architects, engineers, and construction tools, materials, and methods that made each structure a reality, as well as the social, political, and/or religious leaders and movements that inspired its creation. Useful maps help the reader locate the nations, cities, streets, and individual structures mentioned in the text; and numerous diagrams and pictures illustrate tools and devices that bring to life various stages of construction. Finally, each volume contains two bibliographies, one for student research, the other listing works the author consulted in compiling the book.

Taken as a whole, these volumes, covering diverse ancient and modern structures, constitute not only a valuable research tool, but also a tribute to the human spirit, a fascinating exploration of the dreams, skills, ingenuity, and dogged determination of the great peoples who shaped history.

IMPORTANT DATES IN THE BUILDING OF THE SPACE SHUTTLE

1972
(January 5) President Richard Nixon instructs NASA to proceed with the design and building of a partially reusable space shuttle, consisting of a reusable orbiter, three reusable main engines, two reusable solid rocket boosters, and one nonreusable external liquid fuel tank.

1972
(March 31): NASA awards the contract for the main engines to North American Rockwell's Rocketdyne Division.

1953
Rocket Scientist Werner von Braun proposes a fully reusable space shuttle, along with a space station, as part of a manned mission to Mars.

1971
The air force agrees to support NASA's proposal of a partially reusable shuttle, helping to overcome Congressional opposition to the shuttle program.

1972
(August 16): NASA awards the contract for the external fuel tank to Martin Marietta Corporation.

1950   **1960**   **1970**   **1975**

1969
President Richard Nixon's Space Task Group endorses the concept of a reusable space shuttle.

1972
(November 3): NASA awards the contract for the solid rocket boosters to Thiokol Chemical Corporation.

The space shuttle Columbia *launches from Cape Canaveral, Florida.*

1971
The federal government implements budget cuts, and NASA abandons plans for a fully reusable shuttle in favor of a partially reusable design.

1973
Tests on the main engines begin.

1973
(July 26): NASA awards the contract for the orbiter to Rockwell's Space Transportation Systems Division.

1979
The second orbiter, *Columbia* (the first designed to fly into space), is completed.

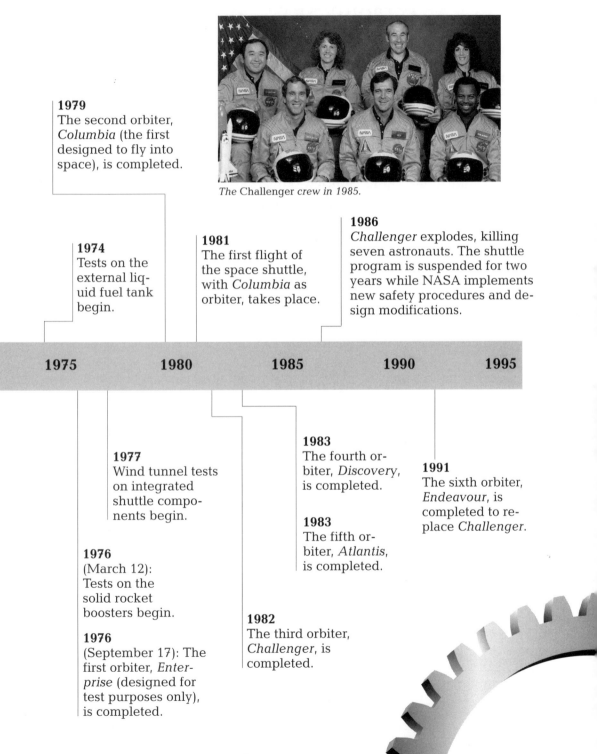
The Challenger *crew in 1985.*

1974
Tests on the external liquid fuel tank begin.

1981
The first flight of the space shuttle, with *Columbia* as orbiter, takes place.

1986
Challenger explodes, killing seven astronauts. The shuttle program is suspended for two years while NASA implements new safety procedures and design modifications.

1975	1980	1985	1990	1995

1977
Wind tunnel tests on integrated shuttle components begin.

1976
(March 12): Tests on the solid rocket boosters begin.

1976
(September 17): The first orbiter, *Enterprise* (designed for test purposes only), is completed.

1982
The third orbiter, *Challenger*, is completed.

1983
The fourth orbiter, *Discovery*, is completed.

1983
The fifth orbiter, *Atlantis*, is completed.

1991
The sixth orbiter, *Endeavour*, is completed to replace *Challenger*.

INTRODUCTION

With more than 2.5 million parts, the space shuttle is the most versatile and enduring craft in the history of space flight. In two decades of operation, it has carried more than 250 people into orbit and traveled more than 340 million miles. It is the first vehicle to enable human beings to live and work in space without cumbersome life-support equipment.

A MAJOR LEAP FORWARD

Before the shuttle, space travel was open to only a few adventurous individuals who were prepared to endure the hardships and risks of exploring what was beyond the earth's atmosphere. The early preshuttle astronauts were highly trained military pilots, many of them battle-hardened, selected because they had demonstrated the physical and mental strength to endure the most extreme and life-threatening conditions.

After the shuttle took its first flight in 1981, however, space became a more friendly and familiar place. Thanks to the relatively secure environment of the craft's crew compartment, the astronaut corps now includes scientists, engineers, and medical doctors—men and women from university and business backgrounds whose training took place in lecture rooms and laboratories, not in military academies. Space journalist Richard Lewis says:

> The Space Shuttle marks the real beginning of space travel. It enables us [for the first time] to function in an extraterrestrial environment, a development of evolutionary scale analogous to the emergence of life from the sea in the far past. . . . It has been suggested that the invention of space travel promises our species immortality, or at least longevity, in that it enables us to tap resources of matter and energy beyond the Earth and thus escape the limits to growth and ultimate survival imposed by the waning resources of one world. The key that unlocks this cosmic larder [supply closet] is a reliable and economically rational system of space transportation. . . . With the development of the reusable Space Shuttle, we have taken the first, long step toward creating such a system.[1]

Despite such praise, the shuttle has also been the target of harsh criticism. To win the federal government's approval on the project, the National Aeronautics and Space Administration (NASA) had to make exaggerated promises. In testimony before a number of congressional committees, the agency's administrators vowed that the shuttle would fly at least fifty times a year, making space travel routine, economical, and safe. As of 2001, however, the vehicle has failed to meet this expectation. Jim Asker, Washington bureau chief for *Aviation Week and Space Technology* magazine, says of the shuttle: "It's a technological marvel, but an economic flop."[2]

The space shuttle has been praised for its technology and criticized for its expense.

DECADES OF OBSTACLES

Congress was wary of the shuttle from the moment of its conception in 1969, and for the eleven years the craft was in development and construction, the nation's elected representatives fought NASA's requests for adequate funding. It was against this background of financial constraint and political hostility that NASA had to overcome the daunting technological obstacles that the shuttle entailed, enduring repeated failures and frustrations.

In the technological realm, the shuttle broke new ground in a number of ways: The orbiter, the part of the system that houses the crew and cargo, was the first space vehicle ever to carry human beings safely to and from space more than once. The orbiter's design required NASA to devise an entirely new way to protect the craft from the intense heat caused by atmospheric friction confronted upon reentry into the earth's atmosphere after each mission. The problems encountered were not completely solved until the very last minute, when the shuttle was sitting on the launch pad awaiting the beginning of the first countdown.

The main engines were the first rocket engines built to withstand more than one flight. During repeated tests, they either failed to ignite or shut down prematurely. On several occasions they exploded, destroying both themselves and the expensive test equipment. Problems with the engines delayed the shuttle program by at least two years and led to cost overruns that, more than once, put the project in jeopardy.

As deadlines were missed, corners were cut. Ultimately, lives were lost. A commission investigating the 1986 *Challenger* explosion, in which seven astronauts died, blamed the disaster on NASA's failure to implement adequate quality control procedures during the shuttle's design and construction. The comission accused the agency of sacrificing crew safety in its unsuccessful attempt to stay within budget.

Although most historians of space flight regard these criticisms as valid, NASA did succeed in building the shuttle against overwhelming odds. Although greatly pared down from the vehicle the agency originally envisioned, at the beginning of the twenty-first century the space shuttle remains the mainstay of America's manned space program. According to Richard Lewis:

The orbiter (pictured) was the first space vehicle to carry passengers safely to and from space more than once.

From years of failure, frustration, and criticism, a reusable space transportation system has emerged. . . . The indifference and doubt that attended the development years have given way to an optimistic perception of this vehicle as a means of opening up a vast new frontier to commercial and industrial enterprise . . . we have created the world's first fleet of true spaceships, an incomparable national resource. In light of its long travail, the accomplishment seems miraculous.[3]

THE BIRTH OF
THE SHUTTLE

The space shuttle had an auspicious beginning. It was conceived in 1969, the year America's Apollo space program had accomplished what many were calling a miracle. On July 20, 1969, Neil Armstrong and Buzz Aldrin became the first human beings to set foot on the moon, and the televised image of the two astronauts planting the American flag on the barren lunar surface was trumpeted as proof that no challenge was too great for the nation's technical expertise. NASA confidently announced that mankind was well on its way to conquering space. The shuttle was expected to be a key element of the next phase of that journey, and NASA administrators, seeking to capitalize on the national space euphoria generated by the moon landing, anticipated that the ambitious project would proceed with clockwork precision to a successful conclusion. That was not to be, however. The space shuttle quickly became mired in political and technical problems that would continue to plague the project for the next decade and would place its future in jeopardy.

THE SPACE RACE

The space shuttle program was the product of the political and technical age in which it was born. Human space exploration began on April 12, 1961, when Soviet cosmonaut Yuri Gagarin, at the controls of the space capsule *Vostok 1*, became the first human being to leave the earth's atmosphere. The event came in the midst of the cold war—the nuclear standoff between the United States and the Soviet Union—and sparked a wave of fear in America.

If the Soviets could put a man in orbit, the U.S. government reasoned, they could also launch all-seeing spy satellites and spacecraft capable of delivering nuclear warheads anywhere on earth. Worse, pride in America's global technological superiority had been bruised, and faith in the supremacy of the free-enterprise economic system had been shaken by the realization that the Soviet Union, a hostile communist nation, had accomplished more than the capitalist, democratic United States. Thus, as a result of the Soviets' accomplishment, cold war tension between the two superpowers, fueled by the belief on both sides that nuclear war was a real and immediate possibility, heightened dramatically.

President John F. Kennedy responded to his concerned nation by vowing that America would be the first to put a man on the moon, and in so doing initiated the space race. In a speech delivered at Rice University on September 12, 1962, Kennedy explained his position:

> No nation which expects to be the leader of other nations can expect to stay behind in this race for space. . . . The eyes of the world now look into space, to the moon and to the planets beyond, and we have vowed that we shall not see it governed by a hostile flag of conquest. . . . We choose to go to the moon in this decade and to do other things, not because they are easy, but because they are hard, because that goal will serve to organize and measure the best of our energies and skills, because that challenge is the one we are willing to accept, one that we are unwilling to postpone, and one which we intend to win.[4]

Kennedy also linked the space program to issues of national defense. In the climate of cold war paranoia, fueled by the belief that the Soviet Union was poised to exploit any perceived American weakness, he had no difficulty in persuading Congress to fund his mission of sending Americans into space. NASA responded

On April 12, 1961, Yuri Gagarin became the first person to leave the earth's atmosphere.

quickly. On May 5, 1961, Alan Shepard became the first American in space and, early in 1962, John Glenn became the first American to orbit the earth. Following Kennedy's assassination in 1963, his presidential successor, Lyndon Johnson, continued the effort, which culminated in the Apollo manned lunar landings. This towering achievement set a lofty standard for NASA, creating inflated expectations for future space ventures, including the shuttle.

MOONSTRUCK

Because the Apollo moon missions occupied such a prominent place in the national agenda, almost overnight NASA grew from a relatively modest body of rocket scientists and space dreamers into a massive federal bureaucracy. In an analysis of the period, the agency's own historians compared Apollo to the Panama Canal: "Only the building of the Panama Canal [a sea passage carved through the Central American country of Panama in

John Glenn was the first American to orbit the earth.

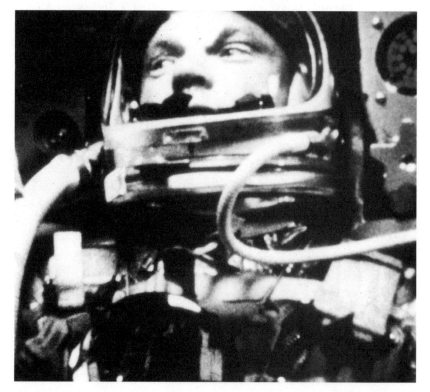

1914 to link the Atlantic and Pacific Oceans] rivaled the Apollo program's size as the largest non-military technological endeavor ever undertaken by the United States."[5]

Between 1960 and 1965, NASA's annual budget burgeoned from $500 million to $5.2 billion, with the largest boost coming after 1962 in response to Kennedy's goal of a manned lunar landing. In 1965 alone, Project Apollo consumed a staggering 5.3 percent of the total federal budget. The final bill for the man-on-the-moon program exceeded $25 billion. Commenting on the astronomical expenditures, veteran space journalist Richard Lewis points out that even though Apollo was a nonmilitary venture, anxiety about national security prompted the government to loosen the purse strings to an unprecedented extent. "These concerns," he writes, "persuaded Congress to write a blank check to finance John F. Kennedy's call for a manned lunar landing within the decade."[6]

For NASA and for the shuttle, this was both a blessing and a curse. In the midst of its greatest triumph, the agency needed a project that would rival Apollo in magnitude. As it turned out, the inspiration for just such an endeavor was close at hand. NASA's most high-profile scientist was Wernher von Braun, head of the agency's Marshall Space Flight Center in Huntsville, Alabama. His success in designing the Saturn V rocket that carried the Apollo astronauts to the moon had cemented his reputation as the world's leading authority on rocket propulsion. Von Braun had a vision of the immediate future of space exploration that dwarfed the accomplishments of Apollo. That vision included the space shuttle.

VON BRAUN'S SPACE UTOPIA

A NASA biography describes Wernher von Braun (1912–1977) as "one of the most important rocket developers and champions of space exploration during the period between the 1930s and the 1970s."[7] As a teenager von Braun became inspired by the idea of manned space travel from reading the science fiction novels of Jules Verne and H. G. Wells. Later, after earning his Ph.D. in aerospace engineering, von Braun became convinced that humankind was capable of travel to the planets and even beyond the solar system. Never lacking in self-esteem, he came to believe that he was the man to lead the way.

In von Braun's expansive concept of space travel, visiting the moon was just a glimmer of more exciting ventures to come. His

President John F. Kennedy (above, left) vowed that an American would be the first to land on the moon. Wernher von Braun (above, right) designed the Saturn V (far right) that took the Apollo crew to the moon.

vision embraced nothing less than a permanent human presence in space. He had been advocating his proposal since 1952, when he authored a groundbreaking series of articles in *Collier's* magazine that bridged the gap between science and science fiction and captured the imagination of the popular publication's huge readership.

By that time he had already established his credentials by building the world's first liquid-fuel rocket and developing the V-2 ballistic missiles with which Germany rained mass destruction on Britain during the Second World War. When von Braun came to America in 1945, the U.S. government was so impressed by his accomplishments that they chose to overlook his association with the German military effort. With his unparalleled credentials, he convinced many people that his goal could be achieved in just a matter of decades. Author David M. Harland says von Braun envisioned a "master plan for space exploration . . . an orderly set of stages aimed at creating a permanent space station serviced from the Earth by a reusable winged vehicle or shuttle, leading to a colony on the moon, and finally a human expedition to Mars."[8]

Von Braun insisted that the technology for the undertaking was almost in place. At a space flight symposium shortly after the *Collier's* articles appeared, he assured an audience of experts that "the United States could launch an artificial Earth-orbiting satellite by 1963, and a 50-man expedition to the moon in late 1964, all at a cost of $4 billion."[9]

Von Braun's projections were grossly underestimated. It took five years longer than he had predicted to put two men temporarily—not fifty permanently—on the moon, and his cost estimates were wildly optimistic. But in 1969 von Braun's seventeen-year-old vision was just what NASA needed to maintain the momentum it had acquired during the Apollo years, and his dream became the stated policy of America's official space agency. Harland says: "These were the ideals that motivated NASA engineers . . . as they pursued the dream of a permanent

ROCKETS AND JETS

The original plan for the shuttle was to equip it with rocket engines, to get it off the ground and into orbit, and jet engines to enable it to land like an airplane. Combustion can take place only in the presence of oxygen. Jet engines use the oxygen present in the atmosphere to burn their fuel. Rocket engines, because they must function outside of the earth's atmosphere, have to carry their own supply of oxygen. That's why the shuttle's large red fuel tank carries both liquid hydrogen and liquid oxygen—the hydrogen is the fuel and the oxygen enables it to burn. The two substances are mixed in the combustion chambers of the three main engines. NASA ultimately decided not to equip the shuttle with jet engines—they were deemed too heavy and too expensive—so the vehicle as it was built must land like a glider and not like an airplane.

presence in space, made sustainable by a reusable winged vehicle providing routine access to space at an affordable price."[10]

NASA GOES TO WASHINGTON

NASA drew on von Braun's vocabulary; fully reusable, routine,
and affordable became almost a mantra for the agency's administrators as they lobbied Congress and the new federal administration of President Richard Nixon to approve funding for
their post-Apollo program. The most easily attainable goal was
the shuttle. To highlight the advantages of a fully reusable
space vehicle, NASA officials intensified their rhetoric, com-
paring the nonreusable Apollo booster and capsule "with operating a railroad and then throwing away the locomotive after
every trip."[11]

NASA commissioned a series of preliminary studies from
various aerospace companies and its own in-house engineers to
explore the feasibility and cost effectiveness of a fully reusable
spacecraft as the first step to attaining von Braun's vision. The
studies indicated that although the concept could be realized,
the price would be high, and the agency used the findings to
mount an intense fund-raising campaign in the nation's capital.
NASA had reason for optimism: Nixon, who was elected president in 1968, was enthusiastic when he welcomed the Apollo
astronauts on their return from the moon. Richard Lewis says:
"Nixon was deeply moved by the event and declaimed that we
had witnessed the greatest week in the history of the world
since the creation [of humankind]."[12]

NASA took Nixon's enthusiasm as an endorsement of
manned space exploration in general. Assuming he would be
as supportive as Kennedy and Johnson had been, Thomas
Paine, whom Nixon's administration appointed to head the
agency in March 1969, presented the president with a com-
prehensive wish list. Historian Joan Lisa Bromberg describes
Paine's bold proposal:

> a space station circling the Earth . . . and another orbit-
> ing the moon. A mixed fleet of shuttles would support
> the stations. Chemically powered shuttles would travel
> from the Earth to the station in Earth orbit while nuclear
> powered vehicles would move between the Earth and
> the moon stations. . . . A lunar base would be established
> and this and the space stations would serve as staging

President Richard Nixon (right), an enthusiastic proponent of space travel, welcomed Apollo astronauts upon their return from the moon.

areas for human exploration of Mars in the 1980s. All this would be funded by a NASA budget of six to 10 billion dollars a year.[13]

NASA's buoyant mood was bolstered in September, when a Space Task Group, established by Nixon and chaired by Vice President Spiro Agnew, presented its report, fully endorsing the agency's vision of the future of the American space program. The report read: "We conclude that NASA has the demonstrated organizational competence and technology base, by virtue of the Apollo successes and other achievements, to carry out a successful program to land on Mars within 15 years."[14] The report went on to recommend the space station and the shuttle as the first elements of this ambitious program, giving NASA everything it wanted and, it appeared, guaranteeing the agency a glowing future in the post-Apollo era.

PROMISES TO KEEP

That future, though, now depended on the agency's ability to build the shuttle quickly and cheaply, as it had promised. However, the technology required to produce a reusable space vehicle had not progressed beyond the theoretical stage. Nobody doubted von Braun, whose specialty was rocket propulsion, when he contended that developing the boosters required to send the shuttle into orbit would not be a problem. However, significant difficulties remained, foremost among them was how to get the craft back on the ground intact and in good enough shape to be used again. All previous spacecraft were single-use vehicles that had parachuted into the ocean after their missions were complete. The shuttle, on the other hand, would have to be piloted to a safe landing, much like a glider.

Another prominent NASA space scientist, Maxime Faget, took on this formidable challenge. Faget, an expert on aerodynamics, had created the wingless capsules of the early manned space programs. However, he was convinced that a winged landing vehicle was within the technical grasp of the aerospace industry and he set about designing preliminary sketches for one. From the start, it became apparent that the cost would be substantial, but Faget stubbornly insisted that it could be done and his determination was a crucial factor in keeping the shuttle program on track during its early stages.

REUSABLE SPACE VEHICLES

The concept of a reusable space vehicle was first seriously proposed in 1933 by German rocket scientist Eugen Sanger. On paper, he designed a vehicle—the Silverbird— a winged space plane that burned a mixture of liquid oxygen and kerosene for fuel and could, he said, fly at ten times the speed of sound, approximately 6,660 miles per hour at an altitude of 100 miles. Unfortunately, the technology did not exist at the time to build such a craft, and until the shuttle was built all space vehicles were wingless vessels designed for one-time use only. When the shuttle finally blasted into orbit in 1981, it flew at an altitude of about 150 miles and a speed of 17,400 miles an hour.

It fell to George Mueller, NASA's associate administrator for Manned Space Flight, to pull the various components of the shuttle project together into a coherent and realistic whole. His job was not to dream big dreams, or to sell the scheme to politicians. Instead, Mueller was charged with the task of coordinating the efforts of the tens of thousands of scientists and engineers, both within NASA and with its partners in the aerospace industry, who would be required to complete the job. Mueller was also the first to realize that the agency had seriously understated the cost of the combined shuttle–space station project in its presentations to Washington. Parts of the program would have to be modified or abandoned entirely unless substantial additional funding could be arranged.

George Mueller, an administrator for NASA, discovered that the agency had set unrealistically low cost estimates for shuttles and space stations.

Thus, say authors Andrew J. Dunbar and Stephen P. Waring, Mueller "began to argue the merits of a Shuttle independent of a space station. Mueller called for a fully reusable low-cost transportation system that might eventually be competitive with other forms of transportation."[15] Mueller anticipated that cost overruns would eventually jeopardize the program as it was originally conceived. He wanted to ensure that if money became a problem, NASA would be able to sever the shuttle from the rest of the plan and justify it as a stand-alone venture, capable of competing with unmanned launch vehicles as a method of putting satellites into orbit.

Mueller was also the first to use the word "shuttle" to describe the vehicle that had been under discussion. While addressing a group of engineers at the Marshall Space Flight Center, he said that NASA needed a "vehicle that's like a shuttle bus."[16] The name caught on, and Marshall's contractors were instructed to use the designation "space shuttle" when referring to the project.

ALL SYSTEMS GO

Mueller's concerns notwithstanding, NASA pushed ahead, asking its allies in the aerospace industry to come up with preliminary designs for the shuttle. The broad parameters were set by Mueller, who, says Richard Lewis, called for "a Shuttle that takes off from Earth, makes rendezvous in orbit [with a space station or other orbiting spacecraft], and returns to Earth with a minimum of control from Earth. It essentially needs only refueling before another trip."[17] Mueller went on to point out that using the shuttle to launch satellites into orbit would be cheaper than the current method, which relied on nonreusable rockets. Lewis notes that NASA was seeking "a reduction in space operating costs from $1,000 a pound for payload [cargo] delivery to orbit by the Saturn V rocket to 'somewhere between $20 and $50 a pound. If we succeed,' Mueller said, 'we can open up a whole new era of space exploration.'"[18]

America's aerospace companies, all eager to participate in another project they expected to be as lucrative as Apollo, responded quickly. Bearing in mind the nonnegotiable requirement of full reusability, most of the companies proposed a two-stage system consisting of two rocket-powered planes. The larger of these two, comparable in size to a Boeing 747 airliner, would be the booster, carrying piggyback style the smaller plane, the orbiter, to hypersonic (five or more times the speed of sound) velocities. The orbiter would then detach, fire its own en-

AN EARLY SHUTTLE DESIGN

Initial plans called for the shuttle to be a fully reusable space plane. Among the many concepts offered by the aerospace industry, the Grumman H-33 was typical. It called for two manned space planes. The larger would take off like a jet with the smaller attached to the top of its fuselage piggyback style. At a designated altitude the smaller plane would detach from the larger, which would be piloted back to earth. The smaller plane would then fire rocket engines to propel itself into orbit. When its mission was over, it would reenter the atmosphere, ignite its jet engines, and land like any other airplane.

gines, and soar beyond the earth's atmosphere into an orbit approximately two hundred miles in altitude. The booster, meanwhile, would fly back to the launch site. On reentry, the orbiter, powered by jet engines, would fly to a predetermined runway once the atmosphere became dense enough to provide sufficient oxygen for the jets to function.

Cost estimates ranged from $5.2 billion to $5.5 billion, and all the proposals required both the booster and the orbiter to carry fuel in internal tanks. Although this worked in theory, it had never been done before, and the onboard fuel tanks would add greatly to the weight of the vehicle. In his book, *The Voyages of Columbia*, Lewis describes these design obstacles:

> The booster had to be a hypersonic aircraft capable of carrying an enormous cargo faster than any aircraft had ever flown before. The orbiter stage had to be capable of controlled flight through the entire range of transonic [both above and below the speed of sound] velocities and manageable in the lower atmosphere at subsonic [below the speed of sound] velocity.[19]

Adelbert Tischler, director of NASA's Office of Advanced Research and Technology, humorously acknowledged both the technical and bureaucratic difficulties that lay ahead when he described the challenge as akin to "putting a ring in the nose of a bull from the wrong end of the bull. But, I'm imbued with the conviction that it can be done if we can work our way through the bull."[20]

By late 1969 the shuttle, or at least a preliminary version of it, existed on paper. NASA had convinced itself that its plan was workable, and that the confident spirit that had succeeded in putting human beings on the moon would see the shuttle through to a successful conclusion.

The Battle to Save the Shuttle

Armed with the endorsement of President Richard Nixon's own Space Task Group, NASA's financial analysts confidently worked out a budget for the full program recommended by the presidential commission, including the shuttle, the space station, and flights to Mars. The numbers were staggering, and, as it turned out, unrealistic.

The Federal Watch Dog

In *Countdown: A History of Space Flight*, T.A. Heppenheimer reports that the shuttle program

> would exceed Apollo both in scope and expense. Its cornerstone would indeed be a manned flight to Mars, at a cost of some $100 billion. In contrast, Apollo had spent a total of $21.35 billion through the first landing on the moon. At a time when the gross national product [a nation's total output of goods and services in a year] was just approaching the level of $1 trillion, the cost of the Mars effort would equal the economic activity of the entire state of California for a full year.[21]

Despite those high estimates, in 1970 enthusiastic NASA administrators presented a formal budget request of $4.5 billion for the first year of the new program to the Office of Management and Budget (OMB), a branch of the federal government that would come to play a major role in the development of the shuttle in the years ahead. According to its official mission statement, the OMB's primary job is to "assist the president in overseeing the preparation of the federal budget and to supervise its administration in Executive Branch agencies. . . . OMB evaluates the effectiveness of agency programs, policies, and

NASA's original budget included the cost of funding the shuttle, a space station, and trips to Mars as shown in this artist's conception.

procedures, assesses competing funding demands among agencies, and sets funding priorities."[22]

In its role as the federal administration's official financial watchdog, the OMB had a shock waiting for the officials at NASA, who, as Heppenheimer points out, "had become accustomed to virtual carte-blanche [generous] treatment during the Apollo years, but [who] now received a cold bath in the Sea of Reality. The OMB rejected the request out of hand, chopping nearly a billion dollars from the proposed budget."[23]

When adjusted for inflation, the cut left NASA with less than half the money it had available in 1966, and it was the first of many financial disappointments to plague the agency during the building of the shuttle. For the first time, the American space program had to justify its expenditures on an item-by-item basis, and that dictated a whole new set of operating procedures that would define both the shuttle itself and the way the groundbreaking research and development required to make the craft a reality would be carried out. The change represented a 180-degree turn from the Apollo era, when limitless funds gave free rein to the agency's creative forces.

WHERE DID ALL THE MONEY GO?

The loss of its privileged position in the hierarchy of national priorities stunned NASA officials. Public affairs analyst Howard E. McCurdy comments:

> It came as quite a shock to NASA professionals when, as the space race came to an end, political considerations began to supersede their professional judgments. This was especially disturbing since the change occurred so quickly after the landing on the moon. To NASA employees, the lunar landings represented the triumph of their organizational philosophy, a professional accomplishment of the first magnitude. Soon thereafter, the space program was forced to enter the political medium, within which most government agencies exist. NASA professionals had to come back to earth.[24]

NASA had never been required to justify itself politically, so the ability of its senior management to assess the mood of the electorate was weak compared to their capacity to judge technical matters. Consequently, the agency's decision makers failed to appreciate how far-reaching were the changes that swept over America during the late 1960s. First, there was a growing bias against technology in general. Programs like the shuttle, says author Richard Lewis,

> were being challenged by the spread of a militant humanism [a belief that human welfare and technological progress were incompatible] among young intellectuals and their [college] campus followers, who considered themselves the vanguard [leadership] of the American cultural revolution. These young militants branded virtually every form of technology, from spaceships to modern agriculture, as antihumanistic.[25]

Second, the federal budget had exploded during Lyndon Johnson's administration. Spending on the Vietnam War and on social welfare programs made enormous demands on the Treasury Department, leaving little money for other endeavors in the years that followed. Lewis observes:

> In the decade between the start of Apollo and the start of the Shuttle, the cost of technological developments had become a critical issue. On the one hand, it was seen as competing with social amelioration [welfare]

Despite the success of the Apollo missions (an astronaut from the last one is seen here on the moon), the U.S. government later severely limited NASA's budget.

MANNEV VERSUS UNMANNED SPACEFLIGHT

A difference of opinion exists in the space community about the relative merits of manned and unmanned vehicles. Unmanned vehicles are cheaper to build and operate because they don't require the safety measures necessary to protect the lives of human astronauts. Many space flight experts feel that unmanned vehicles are the most efficient way to put satellites into orbit. But only a manned vehicle like the shuttle can repair damaged satellites, either while they are on orbit or by capturing and bringing them back to earth. This ability saved the multi-billion-dollar Hubble Space Telescope. Also, without the shuttle and the astronauts it carries, the International Space Station could never have been built. Thus, although members of the space community recognize the benefits of each (manned and unmanned space travel), the debate over which is preferable continues.

programs—taking bread from the poor. On the other, the need for or "relevance" of appropriating billions for a Space Shuttle was fully acceptable only to a technically minded minority.[26]

But even that technically informed minority had its doubts about the value of the shuttle project. A substantial number of space scientists did not share NASA's vision of the supremacy of manned flight in the national space program, fearing that it would siphon funds from unmanned space exploration, which many experts felt could accomplish most of the shuttle's goals more cheaply and with no risk to human life. Representing this point of view, and looking back on the shuttle from the perspective of 1981, Professor James A. Van Allen of the University of Iowa declared in an editorial in the journal *Science*: "It is almost impossible to obtain a go-ahead for a new [unmanned] scientific mission in space. . . . It is time to recognize that the dominant element of our predicament is the massive national commitment of the past decade to the development of the Space Shuttle and the continuation of manned space flight."[27]

THE POLITICAL BATTLEGROUND

Surprisingly, NASA's most daunting opponent was the new president. Despite Nixon's early enthusiasm for space exploration, as a Republican, he did not share his Democratic adversary Lyndon Johnson's allegiance to manned space flight. Says author Joan Lisa Bromberg: "Nixon identified the great manned space programs of the 1960s with his Democratic rivals: in the first instance with John F. Kennedy, who had defeated him in the 1960 presidential election, and after that with Kennedy's successor, [Lyndon] Johnson."[28]

Furthermore, the climate of the country also helped to dampen Nixon's zeal. As Bromberg points out:

> The budgetary squeeze brought about by the [Vietnam] war could make it all the easier for Nixon to cut back on NASA programs. So too could the fact that in the country at large enthusiasm for large technological problems was diminishing. Despite its spectacular performance, NASA . . . was just another federal bureau, reduced to scrapping with other government agencies for its share of a restricted budget.[29]

In the House of Representatives, Congressman Joseph Karth, a Democrat from Minnesota, who sat on the Committee on Science and Astronautics, which played a major role in setting the legislative agenda with respect to space exploration, also reflected the new mood of the country. He mounted a bid to halt NASA's plans. Even though he had previously been an ardent supporter of space exploration, Karth sponsored an amendment to cut off funding for the shuttle and its related programs. The House vote on the bill, in April 1970, ended in a tie, failing to pass, but left the shuttle program extremely vulnerable to future challenges.

By far the most vocal critic of NASA in the post-Apollo years was Senator Walter Mondale, another Minnesota Democrat, who mounted an eloquent campaign against the agency's plans. In one impassioned speech, he said: "I believe it would be unconscionable to embark on a project of such staggering cost when many of our citizens are malnourished, when our rivers and lakes are polluted, and when our cities and rural areas are dying. What are our values? What do we think is more important?"[30] Three months after Karth narrowly failed in the House

Walter Mondale opposed NASA's expensive budget and wanted the government to focus on domestic issues.

to halt the shuttle program, Mondale tried to do the same in the Senate. His bill came within five votes of passing.

In the face of this opposition, NASA lobbied vigorously in defense of its program, even canceling some of the remaining Apollo missions to create room in its budget. Its pleas were ignored by politicians, especially the more liberal elements of the Democratic Party, who felt that the federal treasury should give priority to more pragmatic demands on the nation's financial resources. "It was no use appealing to Congress," Richard Lewis writes in *The Voyages of Columbia*. "Among a number of Democrats in Congress, there was a growing sentiment that the shuttle was out of context in terms of national priorities."[31]

NASA BUCKLES

With the backing of most congressional Democrats assured, Nixon acted on the OMB's recommendations and rejected the proposals of his own Space Task Group, which had advocated building the shuttle, a space station, and a permanent base on the moon as part of a program that would send human beings to Mars. NASA administrator Thomas Paine, finally accepting the magnitude of the opposition he faced, responded with the first of a series of compromises: He abandoned the lunar base and the voyages to Mars. The future of the American space program was now reduced to the shuttle and the space station it was meant to service. "NASA began to lower its sights," says T. A. Heppenheimer. "Still, with these projects [the shuttle and the space station] as a focus for future plans, NASA still had a basis for hope."[32]

Paine pushed the shuttle and the space station as interdependent projects, insisting that one made no sense without the other, with a combined price tag of $10 billion. But even that did not quell the critics, including Mondale and Karth, who vowed to block the money Paine was requesting. NASA was in a difficult situation, but, Heppenheimer says,

> Paine and his colleagues had one more card to play. Their . . . space station could not operate without the Shuttle, but the Shuttle might stand on its own, as an all purpose launch vehicle. By offering low launch costs, lower by far than any expendable booster, the Shuttle might make these expendables obsolete and go on to carry every future payload. It would do this by introducing airplane-like reusability as its route to low-cost launches, thus becoming all things to all people. This prospect might yet allow NASA to proceed with the Shuttle as its next major manned program, and to maintain itself as the agency it had become and wanted badly to remain."[33]

The plans for this lunar base had to be abandoned after NASA's funding was greatly reduced.

ROCKET PROPULSION

Rocket engines generate thrust by expelling combustible gas through a nozzle. According to the third law of motion devised by seventeenth-century physicist Isaac Newton, for every action there is an equal and opposite reaction. Thus, as the shuttle sits on the launch pad, the gas expelled downward from its main engines and solid rocket boosters generates a force pushing upward that lifts the vehicle off the ground and into orbit. For every pound the shuttle weighs, an additional pound of thrust is needed to accomplish that task. That's why the shuttle was designed to be as light as possible—the less it weighs, the more payload it can carry into space with the same amount of rocket propulsion and, hence, the same amount of fuel.

In taking this position, Paine created a situation where the sole justification for the shuttle was financial. No longer could the argument be used that the shuttle was part of a larger whole. It had to stand on its own and fulfill the promises made on its behalf. It had to be so inexpensive to build—and to operate—that government, science, and industry would accept it as the most efficient gateway to the world beyond the earth's atmosphere. It was a bold commitment, full of risk because NASA was basing its promises on cost projections that had not been fully confirmed, but it kept the shuttle program alive.

Yet even that was not enough to silence the critics, especially those in the OMB, who warned they would approve no more than $5 billion for the entire shuttle project. The best numbers NASA could come up with hovered around $7.5 billion, even without the space station. Paine realized that NASA needed allies if it were going to save the shuttle and, because the shuttle was the only major project left, the agency itself.

THE AIR FORCE TO THE RESCUE

Paine made a last-ditch bid to win White House approval for the shuttle in the fall of 1970. He was curtly rebuffed and resigned shortly thereafter. George Mueller, the agency's principal shuttle advocate, had left earlier, leaving the beleaguered program desperately in need of leadership. However, before stepping down, Paine reached out to the air force for support. He and air force secretary Robert Seamans formed the joint USAF-NASA shuttle Coordination Board to explore whether the shuttle could meet Department of Defense needs for launching surveillance satellites into orbit and other military space projects. The alliance was continued by James C. Fletcher, Paine's replacement as NASA's

To garner support for the shuttle program, NASA enlisted the aid of the U.S. Air Force, which was using rockets like this one to launch its surveillance satellites.

administrator, and it would have a profound impact on the final design of the shuttle.

Fletcher determined that to ensure that the shuttle won the approval of the Nixon administration, he would have to get the air force to commit to using the shuttle to launch all its military satellites. It was a hard sell. The air force maintained its own staff of space experts and rocket engineers, and they favored using unmanned boosters to lift satellites into orbit, arguing that they were not only cheaper, but they also avoided risking human life. Seamans's initial response to the NASA overtures was lukewarm at best. He declared that he saw "no pressing need" for the shuttle, and would go only so far as to say it was "a capability the Air Force would like to have."[34]

Seamans also knew NASA desperately needed his support. However, T. A. Heppenheimer says,

> with NASA's space shuttle on the hot seat, Air Force Secretary Seamans found himself in an unusual and quite fortunate situation. NASA needed his service's business more than he needed a new launch vehicle. . . . This put him in a position to drive a very hard bargain. According to the ultimate agreement, the Air Force would contribute its political support and would agree to use the shuttle to launch its payloads. However, it would put up no money for the shuttle's development. . . . In addition NASA would design the shuttle to meet the Air Force's specific requirements."[35]

The terms for air force support were steep, and NASA debated them for months before assenting. Finally, late in 1971 the air force informed both the White House and Congress that it had decided to support the shuttle program and use the shuttle rather than expendable, unmanned boosters to launch its satellites. NASA had overcome another obstacle, but at a price that sparked an internal feud and pitted some of the agency's most powerful forces against each other, necessitating another important compromise in the shuttle's design.

THE DESIGN DEBATES

While political and financial turmoil swirled around the shuttle program, Maxime Faget and other NASA engineers continued to work feverishly on the design of a fully reusable space vehicle. Their proposal called for a first stage, which consisted of a piloted rocket plane booster that would carry the orbiter to supersonic speeds before launching it into space. The booster plane would then return to earth and land like the airplane it essentially was. The orbiter was to have short, straight wings, which would allow it to enter the atmosphere on reentry at a gentle angle, minimizing both structural stress and the heating caused by atmospheric friction. These characteristics would have made the shuttle lighter, requiring less boosting power, and cheaper and faster to refurbish for subsequent flights.

However, Faget's design was capable of lifting a payload—a cargo—of only 12,500 pounds, and it had limited cross-range, the ability to fly to the left or right of its primary flight path after it had reentered the atmosphere. As part of its deal with NASA, the air force had demanded a payload capability of 65,000 pounds and considerably more cross-range than Faget's plans allowed. The first requirement put full reusability in question, and the second forced NASA to consider a delta-wing design (wings that form a solid V shape, extending from the middle to the aft—or rear—of the vehicle's fuselage). Both requirements meant that the returning shuttle would have to enter the atmosphere at a much steeper angle than Faget's design, requiring increased heat insulation and making landings far more difficult and dangerous.

Faget fought fiercely for his design, called the DC-3, pitting him against von Braun's Marshall Space Flight Center, whose engineers insisted they could build rockets powerful enough to propel the bigger, heavier shuttle the air force wanted. NASA had no choice but to accommodate the air force. However, to placate the influential Faget, NASA included a request for models based on the DC-3 concept when it invited proposals from the aerospace industry for specific blueprints for the new vehicle. "It was," says Joan Lisa Bromberg, "an illustration of how hard it was to close a debate within a complex organization like NASA, as well as the clout Max Faget had within the agency."[36]

THE COST OF COMPROMISE

As 1971 drew to a close, the space shuttle program was alive—barely—but only because NASA had agreed to compromise its original vision in several important ways. First, its managers had given up the dream of establishing a moon base and landing a person on Mars. Second, NASA jettisoned the space station from their plans. Finally, in order to win the political support they needed to get any kind of post-Apollo project off the ground, they had allowed the air force to dictate the size and shape of the shuttle orbiter.

By 1972, NASA had agreed to let the air force mandate the shape and size of the shuttle orbiter.

PRESIDENT RICHARD NIXON APPROVES THE SHUTTLE

On January 5, 1972, President Richard Nixon gave his official endorsement to build the space shuttle. His proclamation—obtained from the NASA History Office—read, in part:

> I have decided today that the United States should proceed at once with the development of an entirely new type of space transportation system designed to help transform the space frontier of the 1970s into familiar territory, easily accessible for human endeavor in the 1980s and 1990s.
>
> This system will center on a space vehicle that can shuttle repeatedly from Earth to orbit and back. It will revolutionize transportation into near space, by routinizing it. It will take the astronomical costs out of astronautics. . . . The new system will differ radically from all existing booster systems, in that most of this new system will be recovered and used again and again—up to 100 times. The resulting economies may bring operating costs down as low as one-tenth of those present launch vehicles.

However, as a result of the air force's support, Congressman Karth changed his position and declared his support for the shuttle. Senator Mondale remained unmoved, even accusing NASA of being more interested in keeping the shuttle program alive than in building a viable and cost-effective space vehicle. "NASA has repeatedly changed its design, purpose and justification, not to meet technological or scientific demands, but to make it politically salable [workable]," Mondale claimed. "NASA desperately wants a multi-billion dollar project and will seek any rationale to justify its development."[37] Nevertheless, the Senate also voted in favor of the shuttle.

Even so, the struggle was not yet over. The OMB demanded to see some hard data to back up NASA's claim that the shuttle would be more economical than unmanned, nonreusable launch vehicles. NASA obliged by promising that the shuttle would be a fully reusable vehicle, able to fly up to sixty times a year and deliver payloads into orbit at less than $100 per pound. Critics later flatly accused NASA executives of premeditated falsification on these points. Public affairs analyst Howard McCurdy, for instance, says an anonymous retired NASA executive told him that promising a "shuttle that was cost-effective when you could calculate on the back of an envelope that it wasn't—I don't understand why you would do a thing like that. . . . That's a lie. You never lie. Lying is wrong. That always gets you in trouble."[38]

As it turned out, the compromises NASA made in order to ensure the survival of the shuttle would create serious trouble down the road, with Congress, with the White House, and with the engineers who had to grapple with the impossible task of translating the unrealistic requirements into reality. But the agency had successfully answered congressional critics and allayed the doubts of skeptics at the OMB, enabling the shuttle to survive.

The Shuttle
Takes Shape

The government had given NASA the proverbial green light to proceed with the shuttle, but with a budget that fell far short of the space agency's expectations. Finalizing the design became a constant, and sometimes bitterly acrimonious, struggle between in-house planners, outside aerospace industry contractors, and Defense Department overseers. All this was carried out under the watchful eye of the OMB, which demanded regular accountings from NASA, and a skeptical Congress, concerned that voters would rebel against any cost overruns.

Farewell to Full Reusability

NASA issued contracts to McDonnell-Douglas Astronautics Company of St. Louis, Missouri, and to North American Rockwell Corporation, of Downey, California, to design a fully reusable Integrated Launch and Reentry Vehicle (ILRV). As described by David Baker in his book *Space Shuttle*, the resulting concept

> required a booster to carry the orbiter nearly 80 kilometers [approximately 50 miles] high before separating and returning to Earth, while the orbiter ignited its engines and accelerated into orbit. Both booster and orbiter would have wings for atmospheric flight, both would carry crew members and both would land on a conventional runway.[39]

The proposal caused alarm because it called for two vehicles that were extremely large and expensive. The booster would have been as big as a 747 jetliner, and the orbiter that it was to carry into space as big as a 707. Yet, the combined system could accommodate a payload of only twenty-four thousand pounds, well below the Defense Department requirement of sixty-five thousand pounds. Worse, the cost was estimated at $10 billion to $12 billion.

The gravity of the situation was brought home in 1971, when the OMB pulled the purse strings even tighter. "The OMB announced that NASA's annual appropriation would be limited to $3.3 billion," writes Tom D. Crouch, senior curator of aeronautics at the Smithsonian Institution's National Air and Space Museum. "NASA would have to fall back on a less expensive configuration."[40]

Worse was yet to come. An economic slowdown, or recession, in America put added pressure on the federal budget, and

NASA proposed that the shuttle's boosters be as large as a 747 jetliner (top) and that the orbiter be the size of a 707 (bottom).

the OMB made further cuts in the amount of money allocated to the shuttle. Finally, in 1972 NASA was granted presidential approval to proceed with designs for a vehicle that would cost no more than $2.6 billion. The Defense Department was uncompromising in its demands for maximum payload and cross-range capability. The only option remaining was to abandon once and for all the hope of a fully reusable craft.

MORE BUDGETARY BATTLES

NASA hired a firm of professional economists, Mathmatica Inc., of Princeton, New Jersey, to run a cost-benefit analysis of the shuttle program. The economists' report recommended that the manned booster be replaced by strap-on external fuel tanks. A shuttle with that design would cost more to operate once it was ready for flight, but it would cost far less to build—and NASA's immediate problem was to bring development and construction costs within the budgetary framework dictated by the OMB.

NASA's director Thomas Paine resigned rather than operate within the stringent financial limitations, and many of the agency's managers and engineers sympathized with his decision. His replacement, James Fletcher, incurred their wrath when he decided that a partially reusable shuttle was the only option left. Fletcher, says T. A. Heppenheimer, "quickly realized that if he were to continue to insist on the two-stage fully reusable design, he might wind up with no Shuttle at all; OMB was that strong. He therefore set out to make peace with OMB by meeting its demands, and if that meant stepping on toes in his own agency, he would do it."[41]

But Fletcher's toughest battle concerned the size of the orbiter's payload bay. The OMB had refused to approve the fifteen-by-sixty-foot bay the Department of Defense had demanded to guarantee the shuttle would be able to launch the full range of satellites it was contemplating, and Fletcher knew that congressional antagonism would heighten if the military withdrew its support. In a series of hard bargaining sessions, he failed to persuade the OMB to budge from its insistence that the payload bay be no larger than fourteen-by-forty-five feet; the smaller bay would slash overall costs considerably.

Fletcher then faced a crisis. Space historian Cliff Lethbridge recalls:

Late in 1971, there was a chance that the Space Shuttle program would be halted for more than a year. NASA had no guarantee that President Nixon would recommend any expenditures for the Space Shuttle in his fiscal year 1973 budget, which ran from July 1, 1972 to June 30,

Economists determined that a detachable external fuel tank (like the large one on the right) would bring down the costs of building the shuttle.

Support for the shuttle program was threatened after the OMB initially re-fused to approve a fifteen-by-sixty-foot payload bay (pictured).

1973. President Nixon was ready to present his fiscal year 1973 budget to the U.S. Congress in early January, 1972. If he did not endorse funds for the Space Shuttle in this budget, the program would have faced a stall until July, 1973 at the earliest.[42]

That would have meant thousands of layoffs among NASA employees and perhaps even the closing of several facilities.

Fletcher knew the only way around the impasse was to win over President Nixon himself. In a face-to-face meeting Fletcher reminded President Nixon of his pride the day he welcomed the Apollo astronauts back from their trip to the moon, and equated the need to build the shuttle with America's ability to be a world leader in technology. Finally, Fletcher showed the president a detailed model of the shuttle that was then under consideration. Nixon, reports T. A. Heppenheimer, "held on to it as if he would not give it back."[43]

Later, when Fletcher met with OMB head George Shultz, prepared for yet another difficult negotiating session, he was told that Nixon had already approved the fifteen-by-sixty-foot bay. The Defense Department was still being supportive, congressional critics were silenced, and the shuttle had been saved from the budgetary scrap heap once again.

THE FINAL DESIGN EVOLVES

To win financial approval for a shuttle of any kind, NASA had to drastically scale back its original design. The concept of a reusable launch vehicle was replaced with one that had the orbiter attached to external fuel tanks. These tanks would drop off once their fuel supply was exhausted, lightening the load on the way to orbit. The new concept was still vague on specifics, but it would be cheaper to build (although more expensive to

SOLID AND LIQUID ROCKET FUELS

The shuttle was the first manned space vehicle to use solid rocket fuel as part of its propulsion system. The pros and cons of solid rocket fuel are simple: On the plus side, it is more efficient than liquid fuel. That is, it delivers more thrusting power per pound. The negative side is that once solid rocket fuel begins to burn, it cannot be extinguished. The shuttle's solid rocket boosters are packed with a gumlike mixture of propellant and oxidizer (a chemical compound that releases the oxygen necessary for combustion to take place) that is ignited and burns from the top of the cylinder to the bottom until it is exhausted.

On the other hand, the flow of liquid fuel burned by the main engines can be controlled, much like the gasoline in a car, enabling the shuttle's computers to vary the amount of thrust provided at various times during the vehicle's ascent to orbit. This throttling ability makes it possible to send the shuttle into space without subjecting the astronauts to the excessive downward pushing forces they would have to endure if only solid rocket fuel were used. For the same reason, liquid fuel engines are safer; they can be shut down entirely in an emergency.

operate). NASA was not happy with that trade-off, but it had no alternative that would not substantially exceed the approved budget.

Now that the limits that budgetary considerations would place on the final design were becoming clear, the various NASA centers began to vie fiercely with each other for the most significant roles in creating the shuttle. In notes compiled into a book after his death, Wernher von Braun described the competition and its outcome:

> Roles in the project were assigned as expected, but not before some jockeying for position. The Johnson Space Center in Houston was named overall project manager. . . . It was also responsible for the orbiter. The Marshall Space Flight Center in Huntsville, Alabama, was placed in charge of the engines, tanks and boosters, plus major vibration testing. Kennedy Space Center at Cape Canaveral, Florida, was named as the launch site, with Vandenberg Air Force Base, California, as the second such site to be developed.[44]

At the same time, NASA sent out a call for modified designs to the aerospace companies with which it had been working. They responded with a flurry of concepts, and by August 1971, no fewer than ten different configurations were still under consideration. The "aerospace companies tried hard to sell NASA on the technologies they knew best," says author Joan Lisa Bromberg. "But OMB pressure, by forcing a redesign of the shuttle, sent NASA scrambling for new ideas and aerospace industry engineers were able to provide many of them."[45]

The aerospace corporations were aware that the stakes were high and that, unlike Apollo, cost would be the principal determining factor in which of them won the contract. Bromberg says:

> Because NASA was counting its pennies, firms would also feel themselves constrained to underbid in order to win. On the other hand, the Shuttle was to be an operating transportation system rather than an Apollo-like one-time spectacular. The firms that got the contracts might therefore look forward to orders for a fleet of shuttles, a fleet that in 1972 was being projected as between five and nine. Beyond that fleet could come upgrades and

second-generation craft. . . . And the Shuttle was going
to be NASA's only manned space initiative of the 1970s.
To keep a hand in, aerospace companies needed to work
on it.[46]

THE BIDDING WAR

Like much else about the shuttle, the bidding among aerospace
competitors for the lucrative contracts to build the craft's various
components was fraught with controversy. Though the companies mounted intense public campaigns, critics charge that the
agency's administrators showed preferential treatment to those
firms it had been working with since the early days of the space
program. Defenders argue that the work that needed to be done
was so specialized and so much on the cutting edge of what was
technically feasible that only a limited number of companies
had the expertise to undertake the task responsibly.

The first contract put up for bids was for the main engines.
Three companies submitted proposals, and the contract was
awarded to North American Rockwell's Rocketdyne division.
Another competitor, United Aircraft's Pratt & Whitney division,
quickly filed a complaint with the federal government's General
Accounting Office (GAO), accusing NASA of ignoring Pratt &
Whitney's eleven-year track record designing engines for the air
force in favor of Rocketdyne's merely paper, or theoretical, proposal. In the spring of 1972, the GAO ruled in favor of NASA's
decision. The contract stipulated twenty-five engines at a total
cost of $450 million.

Next was the estimated $3 billion contract to build the or-
biters, the part of the system that would provide astronauts with
a safe environment in which to live and work in space. Four
companies submitted bids, but North American Rockwell once
again emerged the winner. Sensitive to the need to share the
workload—and the payday—among several companies, NASA
attempted to remedy the emerging bias in favor of Rockwell by
awarding subcontracts to those companies that had lost out.
"Grumman got the wing of the orbiter, General Dynamics-
Convair the mid-fusilage, McDonnell Douglas the system that
allowed the orbiter to maneuver while in orbit, and Fairchild In-
dustries the vertical tail fin,"[47] Bromberg reports. By subdividing
the work, NASA created a monumental coordinating and sched-

uling task. Critics later charged that this was a major reason for the cost overruns that plagued the program over the next decade.

Nevertheless, the number of companies involved expanded when the final two contracts were awarded. In August 1973, Martin Marietta was given the responsibility of building the external liquid fuel tank, another controversial decision many historians say was made to favor NASA's friends in the private

North American Rockwell won the bid to build the space shuttle's main engines (pictured).

sector. "According to the company's official historian, Martin Marietta had not even planned to bid," one writer says. "It was NASA administrator [James] Fletcher who encouraged the president of the Aerospace Division to 'go after it.'"[48]

The final major contract, for the solid rocket boosters, went to the Utah-based Thiokol Chemical Corporation in November 1973. This, too, raised red flags among critics of the shuttle program. Fletcher was from Utah, and the state's political leadership

When the Thiokol Chemical Corporation received the contract to provide solid rocket boosters (pictured), some shuttle critics worried that NASA officials were playing favorites.

had lobbied him to give the contract to Thiokol. Lockheed, an aerospace firm that had been excluded from the major contracts, was also given a significant role to play in the shuttle program when NASA asked it to design and build the heat-resistant ceramic material to protect the orbiter from the high temperatures it would encounter on reentry. Bromberg sums up this important phase of the shuttle's development, saying,

> Much of the industry did get contracts. On the one hand, this reflects the fight that the companies mounted . . . the firms fought with every weapon: low bids, appeals to Congress, litigation, advertisements, and the lobbying of NASA officials. On the other hand the evidence, and particularly the subcontracting, also suggests that NASA wanted to spread employment across firms and across regions, and thereby [please] Congress, the administration, and the industry.[49]

THE ORBITER AND THE MAIN ENGINES

Even at this stage, the shuttle's design was not final—modifications would continue to be made even after construction had started. Nevertheless, the basic configuration was fixed. The total assembly—the orbiter and the main engines along with the external liquid fuel tank and the two solid rocket boosters attached to it—stands 184.2 feet tall and weighs 4.5 million pounds at liftoff.

The orbiter itself is 122.17 feet long, 57.67 feet high as it sits on the runway with its landing gear deployed, and has a wingspan of 78.06 feet. The cockpit, living quarters, and experiment operator's station are located in the forward fuselage. This crew module, as it is called, is 2,325 cubic feet in capacity (equivalent to a small bedroom) and is designed to accommodate up to seven astronauts, although it can hold ten in an emergency. In NASA's description, the two-seat cockpit, or flight deck, "is designed in the usual pilot-copilot arrangement, which permits the vehicle to be piloted from either seat and permits one-man emergency return. Each seat has manual flight controls. . . . More than 2,020 separate displays and controls are located on the flight deck."[50]

Below the flight deck, the middeck contains the crew's sleeping quarters, stowage facilities, the kitchen, and the toilet. There

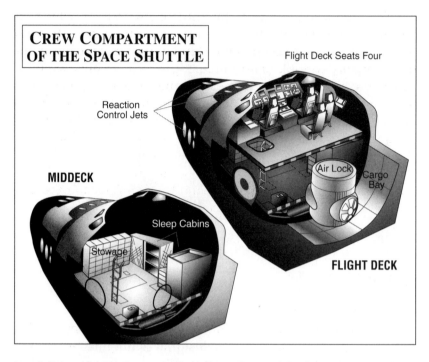

CREW COMPARTMENT OF THE SPACE SHUTTLE

Flight Deck Seats Four

Reaction Control Jets

MIDDECK

Air Lock

Cargo Bay

Sleep Cabins

Stowage

FLIGHT DECK

is additional stowage space below the middeck, where much of the shuttle's complex electronic equipment is also located. The crew compartment has eleven windows, including two that look into the payload bay. Each window is made up of three panes and, at a total depth of nearly three inches, they are the thickest ever designed for see-through flight applications. The crew area is pressurized at 14.7 pounds per square inch and its atmosphere is composed of 80 percent nitrogen and 20 percent oxygen, almost exactly like that on the surface of earth, enabling the astronauts to live and work without wearing special gear.

The crew compartment is joined to the payload bay in the midfuselage of the orbiter by an air lock, which also allows space-walking astronauts to leave the vehicle. The payload bay contains the Remote Manipulator System (RMS), a robotic arm that is controlled from the flight deck and assists in extravehicular activities. Its two doors each contain four heat-radiating panels that keep the shuttle's internal temperature within livable limits. These doors are always open when the shuttle is on orbit, and to ensure that solar heat will not interfere with the panels' temperature-control function, the normal orbital position of the shuttle is with its top turned toward earth and away from the sun.

The aft fuselage contains the three main engines, which assist the solid rocket boosters in getting the shuttle into orbit, and a series of small orbital maneuvering and reaction control engines enable the obiter to make flight adjustments while in space. Each of the main engines, the most advanced rocket engines ever built, is 14 feet tall, 7.5 feet wide, and weighs 3,039 pounds. The engines burn a combination of liquid hydrogen and oxygen to produce 375,000 pounds of thrust apiece. They burn their fuel at a rate that would drain an average-size swimming pool in fewer than ten seconds and generate more than 37 million horsepower. They can withstand temperatures in excess of 6,000 degrees Fahrenheit, and can be throttled, or regulated, so that the astronauts never have to endure acceleration that subjects them to more than three times the gravitational force experienced at earth's surface, a force less than that experienced on many roller-coasters.

THE EXTERNAL TANK AND SOLID ROCKET BOOSTERS

The external tank (ET) is the shuttle's gas tank, containing a little more than 1.5 million pounds of liquid hydrogen and oxygen at liftoff. It is also the only component that is not reused. Approximately eight and a half minutes after liftoff, when its propellants have been fully consumed by the main engines, the ET is jettisoned and breaks up in the atmosphere. The ET is the backbone of the shuttle system, providing structural support for both the orbiter and the solid rocket boosters until they reach an altitude of about 28 miles, when the solid rocket boosters fall away. It continues with the orbiter until just before orbital velocity of approximately 17,400 miles per hour is reached, at an altitude of about 70 miles.

The tank itself is 154 feet long and weighs 78,100 pounds. It contains separate compartments for the liquid hydrogen and oxygen. It is insulated with an inch-thick coating of spray-on foam to keep the propellants cold enough so that they won't vaporize before they are combined. The tank also contains a chamber where the fuels are mixed, and a complicated electronically controlled feed system to make sure they get to the main engines safely.

For the first two minutes of flight, the two solid rocket boosters (SRBs) burn simultaneously with the main engines to provide the extra thrust the shuttle needs to escape earth's

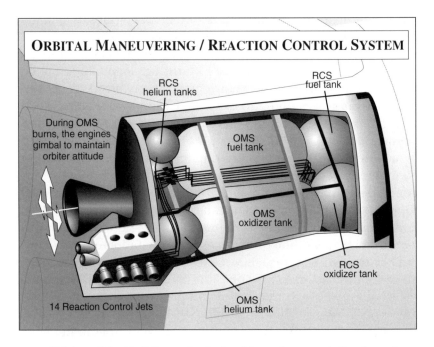

ORBITAL MANEUVERING / REACTION CONTROL SYSTEM

RCS helium tanks

RCS fuel tank

During OMS burns, the engines gimbal to maintain orbiter attitude

OMS fuel tank

OMS oxidizer tank

RCS oxidizer tank

14 Reaction Control Jets

OMS helium tank

gravitational pull. When their fuel is exhausted the boosters detach and parachute into the ocean, where they are recovered by ships, returned to land, and refurbished for further use. They are 149 feet long and, empty, they weigh 87,543 pounds each. The rocket boosters contain a total of 2.2 million pounds of solid rocket fuel, which contributes an additional 5.3 million pounds of thrust already provided by the main engines.

The boosters burn a fuel—which has the consistency of a hard rubber pencil eraser—made from powdered aluminum mixed with an inflammatory substance called ammonium perchlorate. Along with their huge motors, each booster contains an ignition system, moveable nozzles, and electronics. The boosters are made in four segments that are bolted together and sealed with rubber O-rings designed to prevent the ignited fuel from escaping anywhere but through the nozzle.

THE EVE OF CONSTRUCTION

The design, except for a few relatively minor modifications, was fixed; the contracts had been awarded; funding was secured. NASA was about to embark on the construction of the space shuttle. Despite the financial battles with Congress and the president, and the abandonment of the full reusability goal, the agency was convinced that it had a viable project.

But for the program to have progressed this far, NASA had downplayed several serious obstacles that now would have to be hurdled during the construction phase. Principal among these were the main engines, which would be the first rocket engines ever required to perform more than once, and the heat-resistant tiles that would protect the orbiter and the astronauts it contained from the very high, dangerous temperatures of reentry. Like the engines, the orbiter itself was the first space vehicle designed to make more than one flight. In each case, new ground had to be broken. In each case, NASA assured its congressional and financial watchdogs that the situation was under control. However, the aerospace contractors were not as confident that the problems could be solved without significant additional funds. After exploring a number of different accounting methods, NASA assured the government that the total bill for a fleet of four space shuttles would not exceeded $5.15 billion.

But a lot had changed since the shuttle was first conceived, including the loss of the space station that it was initially meant to serve. Unable to justify the shuttle as an element of an enterprise

THE SHUTTLE AS A GLIDER

It is difficult to imagine an airborne vehicle weighing almost a quarter of a million pounds functioning aerodynamically like a lightweight glider, yet that is exactly what the shuttle does after it reenters earth's atmosphere. In space, the shape of the shuttle doesn't matter because it encounters no air resistance. But as the craft plummets through the atmosphere on reentry, it encounters increasingly strong air resistance the closer it flies to earth. These air currents could cause the orbiter to break apart were it not designed to move through them efficiently. However, the orbiter's elongated structure and delta (V-shaped) wing permit it to navigate in the atmosphere like an airplane. Unlike a plane, though, it is not propelled by jet engines. The only force operating it is the force of gravity. So, like a glider, the only way to control its descent is through its aerodynamic design.

ATMOSPHERIC FRICTION

When the shuttle reenters the atmosphere after a mission, it is traveling at approximately 17,400 miles an hour at an altitude of about 150 miles. There is no single point at which the atmosphere ends; it just gets thinner and thinner the farther above earth it is measured. Conversely, as the shuttle falls toward earth on its descent, the atmosphere it encounters becomes thicker. That means that more air molecules per cubic foot are present. Friction results when the air molecules collide with the outer surface of the orbiter. This slows the shuttle down, but it also generates heat so intense that without some sort of protection it would melt the orbiter's aluminum fuselage. The heat-resistant tiles prevent this by absorbing the impact of the air molecules that strike the shuttle, thus permitting both vehicle and crew to land safely.

Workers apply heat-resistant tiles to the shuttle so its fuselage will not melt upon reentering the earth's atmosphere.

that would culminate with human visits to other planets, the agency took the approach that the shuttle would make access to space cheaper, both for the military and for industry. Says Joan Lisa Bromberg:

> There was a profound difference between building a Shuttle to service a space station as part of the saga of human exploration of space, and building a Shuttle that had no space station to go to. NASA had come to stress the economic payoff: the Shuttle would be a cost-effective system for placing (or retrieving or servicing) military, scientific, and commercial satellites. . . . A large number of missions—provided, of course, that the cargo bay was full on each of them—would allow the agency to amortize [spread out] the cost of the Shuttle over many missions and thus justify its claim that the Shuttle would make access to space cheaper.[51]

4

LESSONS IN REALITY

Although the administration had approved the construction of the space shuttle, critics of the project continued to be vocal in their opposition. Senate Democrats, led by Maine's Edmund Muskie, Edward Kennedy of Massachusetts, and the still-skeptical Walter Mondale accused President Nixon and NASA of playing pork-barrel politics—meaning they believed the agency used the lucrative shuttle contracts to curry favor with powerful aerospace companies. Author Richard Lewis speculates:

> The more NASA sought to emphasize the cost-effective benefit of the Shuttle, the more convinced its opponents became that it was a make-work [lucrative but impractical] scheme of the military-industrial complex . . . the notion that space activity diverted funds from the human needs influenced the final decision to build the Shuttle on the cheap.[52]

The outcome of that underfunding, Lewis concludes, is that the shuttle was "squeezed and pulled into final shape by economic and social pressure."[53]

Yet, NASA continued to insist that construction problems would be minor. Dale Myers, associate administrator for shuttle development, assured Congress that there were no major problems left to be overcome and that the shuttle would be operational no later than 1978 or 1979. "No technological breakthroughs are required," Myers maintained. "Although significant innovation is required, principally to keep development costs low, the basis for the technology required is solidly in hand."[54]

Despite Myers's optimistic assertion, significant problems began to appear almost immediately. The most difficult of these concerned the main engines, the heat-resistant tiles, and the complicated computerized navigational hardware and software.

NASA's strategy called for work to be done simultaneously on all components of the shuttle, with increasing degrees of integration until the entire system was up and running. Consequently, when any of the components presented a difficulty, work on the others had to cease until the problem was solved.

The agency's planners realized during the design phase that the main engines would be difficult to build, and therefore decided to designate this element as the component around which all the others would be scheduled. Any delays in the manufacture and testing of the main engines would have serious ramifications, including cost overruns and layoffs, for the entire program and the more than fifty thousand people who were involved in translating the shuttle from blueprint to reality. There were many delays. Each day lost drove up costs, gave shuttle opponents more data on which to base their attacks, and necessitated additional money-saving compromises. Some of these compromises, critics charge, were so extreme that they eroded the safety of the vehicle.

Despite NASA's insistence that problems would be minor, significant difficulties occurred with the construction of the shuttle.

ENGINE TROUBLE

Richard Lewis describes the building of the shuttle's main engines as

> one of the toughest jobs NASA and the contractor, Rocketdyne, had ever tackled. The Shuttle directorate underestimated the order of difficulty of developing this advanced, high-pressure engine, but once having started, it had to keep going. . . . Engine troubles were to continue to delay and threaten the safety of the program right up to launch.[55]

The crux of the problem was that even though the engines were meant to be reused, they were also called upon to perform

The main engines on Discovery *(pictured) were the first ones designed to be used successfully more than fifty times.*

at a higher level than any of their predecessors each time they were fired up. Lewis says:

> The SSME [Space Shuttle Main Engine] is the first rocket engine I know of that is designed to be reused 50 times or more, in an airline style operation. All other liquid-fuel rocket engines and solid-fuel rocket motors are fired only once and then thrown away. Reusability alone required a higher order of technology than had been achieved before in the space industry. In addition, the SSME would operate at higher pressure and for longer periods than any previous American main engine.[56]

The litany of difficulties that created major delays included "14 engine test failures . . . caused by faulty seals, uneven bearing loads, cracked turbine blades, cracked fuel injector posts, heat exchanger malfunctions, value breakdowns, and ruptures in the hydrogen lines,"[57] Lewis points out. Eight of the failures resulted in devastating fires that damaged the engines, the test equipment, and even the aft fuselage of the orbiter mock-up employed in the testing process. The destruction caused by the fires often made it difficult to identify the cause and required NASA to double the number of engines it had originally ordered.

These failures unfolded over a three-year period, from 1974 to 1977, casting serious doubt that NASA would be able to make good on its promise to launch the shuttle by 1978. The program was dealt a further blow in 1977, when the OMB imposed more than $300 million in additional budget cuts. The unexpected financial crisis forced NASA to cut yet more corners, leading to further delays in building the main engines. Says T. A. Heppenheimer:

> In response, NASA instituted a strategy whose very title "success-oriented management," suggested hope but concealed desperation. In contrast to Murphy's Law, that anything that can go wrong will go wrong, this approach assumed that everything would go right. In the words of an official, "It means you design everything to cost and then pray." . . . For the SSME, it meant testing full-scale engines before their major components were proven and reliable.[58]

LIVING IN SPACE

Human beings cannot live in space. To survive, they must replicate conditions that exist on earth. Such replication is the function of the crew compartment of the shuttle's orbiter. The crew compartment design creates an atmospheric pressure of 14.7 pounds per square inch, the same as at sea level. The air that astronauts breathe contains about 20 percent oxygen, the same as earth's atmosphere. Temperature remains at approximately 70 degrees Farenheit because the heat generated by the crew's bodies and the orbiter's electronics radiates into space through panels on the inside of the payload bay doors. Electricity is provided by fuel cells that use hydrogen and oxygen and give off pure water as a byproduct of the reaction. The crew then uses this water to drink and bathe. In this way, the orbiter creates earthlike conditions that can support seven people for up to fourteen days.

MISSED DEADLINE

In 1972 NASA promised that the shuttle would make its first flight in March 1978. "But," says Heppenheimer, "when that month arrived . . . the engines were in no condition to do much of anything."[59] While a new schedule for the first launch was being worked out, a major engine failure occurred at the end of December 1978, and two others failed in May and July 1979. In response, a report by a committee of the National Academy of Sciences accused NASA of seriously underestimating both the technical problems and the cost of developing the SSMEs.

Administrators pushed ahead with the testing of the main engines, relying on the belief that the can-do philosophy that had enabled them to put a man on the moon would see them through this difficult period. But, this time, funds at their disposal were far more limited, government oversight of their spending much more stringent, and their foes in the Senate and House of Representatives far more vocal. Many historians of space flight feel that NASA disguised the magnitude of the problems plaguing the main engines in an effort to keep the project going. "Some critics believed that potentially dangerous flaws were being overlooked in the rush to build and fly a new spacecraft with ever shrinking resources,"[60] says author Tom D. Crouch.

The three main engines would be required to fire for a total of eight minutes when the shuttle was launched. By the end of 1979, however, a year following the deadline to which NASA had committed, the agency was far from that goal. It was not until early 1980 that the first successful ground test of an individual engine was accomplished. NASA was jubilant, and established March 1981 as a new target date for the first shuttle flight.

But the problems with the SSMEs had still not been resolved. On April 16, 1980, temperature limits on one of the fuel pumps were exceeded and the engines shut down after just 4.6 seconds of what was scheduled to be an eight-minute test. A second successful test was followed immediately by another failure.

Only in January 1981 did all three engines perform successfully for the required time. It was clear that NASA was going to miss another deadline for getting the shuttle off the ground. Cost overruns were mounting, and critics once again were clamoring for the cancellation of the program. Luckily for the shuttle, however, Jimmy Carter, who had been elected president in 1976, was a strong NASA supporter. In 1978 he issued a presidential directive endorsing the shuttle, and in 1979 made it clear to Congress that he wanted enough money in the budget to guarantee that the program would be completed.

President Jimmy Carter wanted money included in the federal budget to complete the shuttle program.

THE EXTERNAL TANK AND THE SOLID ROCKET BOOSTERS

In contrast to the main engines, progress on the external liquid fuel tank and the solid rocket boosters went relatively smoothly. Nevertheless, both broke new ground in space technology and both encountered serious obstacles. Weight, the force that gravity exerts on an object, is a crucial consideration for any space vehicle. The heavier a spacecraft is, the greater the propulsive power needed to lift it into orbit. That, in turn, requires more fuel, which adds to the total weight, and so on. NASA was therefore on a constant mission to find ways to reduce the shuttle's weight.

The aerospace company Martin Marietta had already begun tests on the external liquid fuel tank when NASA told its engineers to reduce the weight by six thousand pounds and cut costs at the same time. (The production price of the tanks was important because they are the only part of the shuttle system that is not reused.) Engineers at Martin Marietta felt they had already pared the weight of the tank as much as possible without impairing its structural integrity. "Any further weight savings was going to require redesigning significant portions of the tank," says author Dennis R. Jenkins. "The dual goals of lowering the ET's [external tank's] weight and also lowering its production costs were often at odds."[61]

Nevertheless, Martin Marietta succeeded in carrying out NASA's wishes. In all, the company achieved a total weight reduction of about ten thousand pounds. However, the modifications took four years and resulted in removing a layer of fire-retardant latex, which made the tank somewhat less safe than the original plans called for but was still within acceptable limits according to the standards NASA adopted.

In contrast with the difficulties encountered with the other shuttle components, the SRBs presented few problems during development and construction. Ironically, they became the most notorious elements of the system when they malfunctioned and caused the *Challenger* explosion in 1986; cost-cutting during their manufacture and testing was blamed for the disaster.

The shuttle was the first manned vehicle to use solid rocket fuel. This had both good and bad consequences. On the plus side, solid fuel burned efficiently and provided huge propulsive power. On the negative side, once solid rocket fuel is ignited nothing can be done to stop the controlled explosion that results. Unlike the liquid-fuel-driven main engines, the SRBs could not be shut down, no matter what the circumstances.

For this reason, NASA and the solid rocket booster manufacturer, Thiokol, planned and carried out extensive tests on all aspects of the boosters. However, Jenkins points out that in order to keep costs down "none of these tests attempted to duplicate the loads or conditions anticipated during ascent. This oversight would gain particular importance after the [*Challenger*] accident."[62] Recovery procedures, on the other hand, including the parachutes that would allow the boosters to descend without damage, were conducted under full flight conditions to enable NASA to stress the point that this component would be reusable.

THE ORBITER AND THE NAVIGATION SYSTEM

The contract for the orbiter called for a fleet of four fully operational vehicles as well as one test vehicle that would be similar to the others except that it would lack engines and navigational

NASA officials monitor the Enterprise's *landing capabilities after it separates from a modified 747 (bottom).*

equipment. The purpose of this orbiter, which was called *Enterprise* after the *Star Trek* space ship (NASA decided on the name after it was deluged with letters from fans of the science fiction television series), was to verify the aerodynamic properties and landing capability of the shuttle in a series of tests during which the craft was carried to an altitude of seventeen thousand feet on the back of a modified Boeing 747 jetliner, released, and piloted back to the ground on a nearby runway.

Originally, the four fully functional orbiters were designed to be equipped with jet engines, as well as rocket engines, to aid with landing after reentering earth's atmosphere. However, it quickly became apparent that the jet engines would add a great deal of weight. Since every extra pound meant one fewer pound of payload, this was not acceptable to the Department of Defense or, consequently, to NASA. Thus, the agency decided to abandon the jet engines.

This decision was made virtually at the last minute, and it was one that significantly compromised the safety of the astronauts because it limited their ability to adjust the flight path during landing if wind currents blew the shuttle off course. Before eliminating the jet engines, NASA described the shuttle as a spacecraft that would take off like a rocket and land like an airplane. After the mission, it became a vehicle that took off like a rocket but landed like a glider—a glider that weighed almost a quarter of a million pounds. This was a problem because it put a great deal of stress on both the heat-resistant tiles and the shuttle's computer navigation system. Richard Lewis explains:

> At reentry, the orbiter would have tremendous kinetic energy [energy generated by its momentum]. . . . Translated in terms of velocity, it amounted to five miles a second. That's 18,000 miles an hour. The space plane had to slow down to 300 mph to land. So most of the energy was transformed into heat as air friction slowed down the vehicle. The heat shielding deflected the heat; otherwise the orbiter would burn like a meteorite. From reentry on, the velocity of the craft was managed by a delicate balance between lift and drag forces [air pressure] in the atmosphere and by the gravitational pull of the Earth. The balance controlled the energy of the descent. [The orbiter] required a superhuman pilot; it re-

quired a computer, or, as in the case of the orbiter, a committee of them.[63]

Although no human pilot could react fast enough to implement the thousands of adjustments that would have to be made to ensure a successful landing, there was a question about whether the computing power available in the mid-1970s was sufficient. Rockwell's contract to build the orbiter included the onboard computers, but NASA decided to hire IBM to develop the software because of its vast experience in designing complex computer systems. Because the computers would play such a vital role, four completely independent systems were installed, each one capable of doing the job if any, or all, of the others failed. As it turned out, the computer system, which involves more than 2 million lines of sophisticated coding, proved itself to be one of the most trouble-free aspects of the shuttle program.

The space shuttle was originally designed to take off like a rocket and land like an airplane, although in the end, it landed like a glider.

THE HEAT TILES

As the main engines were falling behind schedule, so, too, was production of the first fully equipped orbiter, *Columbia*. The problem was not the groundbreaking computer system. Nor was it the integrity of the crew compartment, or the functionality of the payload bay. All of these fundamental components were under control. The difficulty was with the heat shield, designed to protect the orbiter from melting during reentry. Consisting of thirty-one thousand ceramic tiles, this thermal protection system was another first for the shuttle. Previous spacecraft were insulated from the intense temperatures of reentry by an ablative coating, a substance that carried heat away from the vessel by burning off as the craft plunged through the atmosphere.

Although ablative coatings did the job they were required to do, they were not acceptable for the shuttle's orbiters, which were required to fly again with a minimum of maintenance and refurbishing between flights. The cost of replacing an ablative shield after each mission would have been prohibitive. Instead, NASA opted to use heat tiles made from fibers of sil-

WHY THE SHUTTLE LAUNCHES FROM FLORIDA

The earth rotates from east to west at a speed that is fastest at the equator and decreases with every degree of latitude north or south of that point. At the latitude of the Kennedy Space Center on Florida's east coast, that speed is about nine hundred miles an hour. The rotation creates a slingshot effect that gives vehicles launched in an easterly direction an extra propulsive boost into space, enabling them to carry larger payloads than if they were launched in any other direction. NASA also has a policy of never launching any spacecraft over land to minimize risk to the civilian population in the event of a catastrophic failure. To maximize the weight of the cargo the shuttle carries, it must take off toward the east. To maximize the effects of the earth's rotation, it must take off as close to the equator as possible. Launching over the Atlantic Ocean from the east coast of southern Florida meets all the criteria.

ica, the extremely hard organic compound that is the principal constituent of sand. The tiles were lightweight, almost entirely impervious to heat, and easy to manufacture. The "resistance of the tiles to temperature change is astonishing," says Richard Lewis. "One can hold a tile by the edge while the center glows red hot and not feel heat."[64] In tests, the tiles were heated to more than twenty-three hundred degrees Fahrenheit, then plunged into cold water without cracking.

But the six-inch and eight-inch tiles (the different sizes are required to sustain the different levels of stress the tiles encounter on reentry) turned out to be much more difficult to install than anyone anticipated at NASA's Ames Research Laboratory or the Lockheed Missiles and Space Company, where they were developed. The twenty-four thousand six-inch tiles, black in color, range in thickness from one to five inches and are placed where the highest reentry heat is encountered. The seven thousand eight-inch tiles are white in color, vary between one and three inches in thickness, and are placed on areas of the shuttle that endure less extreme heat.

If any tile protruded higher from the surface of the obiter than those around it, it would be torn off by the extreme atmospheric forces encountered on blastoff and landing. Therefore, each was assigned a specific location on the orbiter's shell. Lewis says:

> Each tile was marked to identify its position on the structure. The "ID" matched its assigned position on the aluminum skin [of the orbiter]. Each tile was packed separately in a plastic shipping container and protected from contamination by a thin plastic wrap. The brittle square could not be touched by bare fingers for fear of leaving a thin smear of oil on the surface that would damage it in a vacuum. All installers wore plastic gloves.[65]

During the installation process, each tile was positioned and checked to make sure it fit properly in its designated place. Then it was repacked and stored until it was needed. It was then glued to a felt pad, which was in turn cemented to the orbiter's skin. The process took thirty-two hours for each tile, and it had to be done one tile at a time.

In March 1979, NASA was committed to an October launch date. To save time, officials decided to move the first orbiter—the

Columbia—from the Rockwell plant in Palmdale, California, to the launch facility at Cape Canaveral, Florida, so that the tiling could be completed while the orbiter and tanks were being assembled. However, they encountered a lack of skilled workers and were forced to hire college students to do the job. Training created further delays and failed to equip the students with the expertise necessary to do the job at maximum efficiency. On average, it took one worker about forty hours to test, install, and inspect one tile—one tile per worker per week. As the launch deadline approached, desperation was running so high that "NASA officials swore that if they could find any other shielding that would be reusable, they would resort to it instantly," Lewis says. "But although headquarters authorized research into alternative thermal protection systems, the [tiles] remained the only one that seemed to be feasible."[66]

LAST-MINUTE GLITCHES

In its rush to meet the October 1979 launch deadline, NASA decided to install the heat tiles on *Columbia* before their durability had been fully tested in wind tunnel trials. Although the heat-resistant qualities of the tiles had been fully demonstrated in the laboratory, it was only after 75 percent of them had been painstakingly installed that wind tunnel data revealed they had a tendency to fall off or split under stress. As a result, thousands of tiles had to be replaced and retested during the fall and winter of 1979, causing NASA to miss its scheduled launch date yet again. "Why had NASA failed to run flight tests of this new shielding on aircraft or reentry vehicles earlier in the development program?" Lewis asks. "I raised this question several times at status briefings. The answer was always the same: it would have been too costly."[67]

NASA had originally planned to include a space walk during *Columbia's* first flight; this would allow the astronauts to fix any damaged tiles while the shuttle was on orbit. The space walk was quickly abandoned as too risky, however, and thousands of additional tiles had to be removed, reinforced, and reinstalled to ensure there would be no significant problems in flight. This, along with continuing problems with the main engines, pushed the date of the first mission back to early 1981. It also forced the agency to appeal to Congress for additional funds. To get the first shuttle off the ground, money that had

WHY THE SHUTTLE ROLLS AFTER LIFTOFF

Shortly after it clears the launch pad tower, the shuttle executes a ninety-degree roll-over maneuver. It does this because NASA, to save money, decided to launch it from pads built for the Saturn V rocket that sent the Apollo astronauts to the moon. Because the shuttle, with its fuel tank and solid rocket boosters, is shaped differently from the long, cylindrical Saturn V, it must sit on the pad with its back pointed in its intended direction of flight. Therefore, as soon as the shuttle rises above the tower, its computers cause it to roll over to orient it to the correct flight path. The extra computer power required to accomplish this tricky maneuver cost far less than modifications to the launch pad.

The shuttle rolls ninety degrees after lift off because as it sits on the launch pad, its back faces the intended direction of flight.

been earmarked for building three future obiters—*Challenger, Discovery,* and *Atlantis*—had to be channeled into completing *Columbia.* In all, an extra $925 million was allocated to *Columbia* at the last minute.

Congress, clamoring over the delays and the cost overruns, accused NASA of mismanaging the shuttle project and hinted that even at this late stage it might cancel the project. Once again, the agency had to turn to the Department of Defense for support. Undersecretary of Defense William Perry testified before a congressional committee that the military was totally dependent on the shuttle for all its projected satellite launches; the program had to continue.

By the summer of 1980, *Columbia's* first flight had been postponed nine times. Problems with the main engines and the heat shield had not been fully resolved, but NASA felt confident enough to predict that the first manned orbital flight of the space shuttle would take place in March 1981. Congress remained skeptical, but because of the Department of Defense's testimony both the Senate and the House of Representatives believed the project had passed the point of no return. Reluctantly, they accepted that the shuttle was going to become a reality, even though the cost now stood at $8.7 billion. The initial estimate of $5.15 billion was a dim memory, and the new figure did not even include the expense of building the three additional orbiters that, along with *Columbia*, would make up the fleet.

5

The Shuttle
Blasts Off—Finally

In December 1980 and January 1981, the main engines completed two successful tests and were pronounced ready for flight. Richard Smith, director of the Kennedy Space Center in Cape Canaveral, Florida, announced that the heat-resistant tiles had been largely installed; the few remaining tiles, he insisted, could be put in place while *Columbia* was on the launch pad. The orbiter was rolled out of the Orbiter Processing Facility, where most of the tile work had been done, and into the huge Vehicle Assembly Building to be attached to the external tank and the solid rocket boosters. The short route between the two buildings was lined with temporary workers who had been laboring night and day to install the tiles. One of them shouted: "Hail, *Columbia*! We who are about to be unemployed salute you!"[68]

During the following two weeks, a series of ground tests was carried out with the two-man crew, Commander John Young and pilot Robert Crippen and their backups Richard Truly and Joe Engle, alternating at the controls. After one such test, Young, a veteran astronaut and test pilot, acknowledged to reporters how different the shuttle was from all previous spacecraft. In previous spacecraft, the pilot could bypass the computer controls and fly manually. With the shuttle's complexity, human reaction time was too slow for that. The pilot can control the shuttle only through the computers. Should the computers fail, Young admitted, the crew would be helpless.

During this crucial time, America elected a new president who in turn appointed NASA a new chief administrator. Following his inauguration on January 20, 1981, Ronald Reagan replaced Jimmy Carter's appointee Robert Frosch with aerospace

THE FIRST SHUTTLE COMMANDER

Columbia's commander, John Young, was a veteran astronaut when he flew the first shuttle mission. In fact, he was gathering rocks on the moon in 1972 as a member of the crew of *Apollo 16* when he received a radio communication from NASA that he would be a pivotal member of the shuttle team. Young, born in 1930, graduated with a degree in aeronautical engineering from Georgia Institute of Technology, and was a navy test pilot before joining the astronaut corps. He made his first space flight on *Gemini 3* in 1965, and commanded a second shuttle mission in 1983. In all, he logged 835 hours in space before becoming chief of the Astronaut Office, the branch of NASA that deals with the selection and training of astronauts. In 1996 he became associate director of the Johnson Space Center in Houston, Texas.

John Young was a very experienced astronaut by the time he commanded the first shuttle mission.

industry executive James Beggs. On the eve of *Columbia*'s liftoff, Frosch held a farewell news conference at the Kennedy Space Center during which, Lewis says, "he expressed regret that he had not tried more vigorously to break out of the OMB's funding constraints. He should have rocked the boat, he said."[69]

LAUNCH DATE APPROACHES

However, the change in administration had no significant effect on the shuttle. The long struggle to build the world's most advanced space vehicle was over at last, and it had been pronounced flight-ready. The massive doors of the Vehicle Assembly Building were raised, and the *Columbia*, mounted on a mobile launch platform, was carried on the back of a giant tractor-trailer three miles to launch pad 91B.

A final test to confirm that *Columbia* was ready for flight was conducted on February 20, 1981. To check out all systems, the main engines were fired for twenty seconds at 100 percent of their power while the shuttle remained anchored to the launch pad. For this crucial test, a three-mile area around the launch pad was cleared, and a beach close to the Kennedy Space Center on the south Florida coast was evacuated. The failure of the main engines at this stage could detonate the shuttle with the force of a three kiloton bomb.

The firing appeared to go without a hitch, but a subsequent test revealed that a portion of the insulation on the external fuel tank had peeled away. The launch date, set for March, had to be postponed while engineers wrestled with the problem. Two other events delayed the launch even further. Workers at Boeing, which was involved in last-minute tinkering with the heat tiles, went on strike and two key Rockwell technicians died of asphyxiation during a hydrogen leak in the shuttle's aft fuselage.

Finally, the liftoff was set for 6:50 A.M. on April 10, 1981. That morning, Young and Crippen were strapped securely into their seats in the cockpit of the orbiter. The countdown proceeded normally until T (launch time) minus nine minutes. At that point, the count was held so that mission control could investigate a discrepancy between the backup computer and the other four. The difference amounted to just forty thousandths of a second, but it was enough to abort the launch.

The space shuttle Columbia *had to pass a series of tests before it could be launched.*

THE BIG DAY

On Sunday, April 12, Young and Crippen awoke before sunrise and ate a breakfast of steak and eggs. They donned their flight suits and boarded a van that transported them to the launch pad. Once there, they rode an elevator 147 feet to the orbiter access arm and entered the cockpit. Behind them, technicians securely bolted the hatch. This time, the computers were properly synchronized and the countdown proceeded without a hitch. However, at T minus fifty seconds the most advanced of the two radar systems that would be used to track the shuttle's ascent into space shut down. Mission control made a hasty decision to go ahead with only the inferior system operating.

At T minus twenty-eight seconds, launch director George Page read Young and Crippen a message from President Reagan and his wife: "for all Americans, Nancy [Reagan] and I thank you. May God bless you."[70] At T minus four seconds, the main engines came to life with a thunderous roar, and an immense cloud of flame billowed from the floor of the launch pad. The solid rocket boosters fired and, as the ground shook, spectators three miles away felt the warm blast of the exhaust. Slowly, the shuttle began to rise from the ground. At

The space shuttle Columbia *lifted off for the first time on April 12, 1981.*

three seconds past 7 A.M., precisely on time, it cleared the launch tower.

At two minutes and twenty seconds, the solid rocket boosters detached from the orbiter and the external tank. *Columbia* was flying at sixty-two hundred feet per second past the west Atlantic island of Bermuda, heading southeast. "What a view, what a view!"[71] shouted Crippen, who was making his first space flight. The monitor attached to his body indicated that his heart was racing at 130 beats per minute. Young, a space veteran, was calmer. His heart was beating at a steady count of 80 per minute.

The main engines cut off at eight minutes and thirty-four seconds, and the external tank separated from the orbiter. At liftoff, the shuttle weighed almost 4.5 million pounds. It now weighed just 214,000 pounds, having burned its fuel and jettisoned the solid rockets and the external tank. Still, it was the heaviest vehicle ever to enter space. After a short firing of its orbital maneuvering system (OMS) engines, it settled into an orbit 132 miles above earth, traveling at 25,670 feet per second. Three more OMS burns raised its altitude to 150 miles and brought it up to orbital speed of approximately 17,400 miles an hour, about six times faster than a high-powered rifle bullet.

CHECK-OUT

Young's and Crippen's first task once they had stabilized the *Columbia*'s orbit was to check out the operation of the payload bay doors, which also contained the orbiter's on-orbit temperature control system. If the doors failed to open, the crew compartment would heat to unbearable levels and the flight would have to be aborted. The two doors each weigh 3,264 pounds on the ground (in space, they are weightless because the shuttle's orbit—essentially a freefall around earth—negates the weight-producing force of gravity), are 60 feet long and 6.7 feet wide. They are connected to the orbiter by thirteen hinges and operated by electric motors that are controlled from the rear of the cockpit.

When Crippen successfully opened the doors, both he and mission control were in for a shock. The first thing he saw as he looked through the window and beyond the payload bay was that heat tiles were missing from the aft fuselage. The sight, relayed back to earth by television cameras, created a momentary panic. A count revealed that a total of 16 tiles were missing and 148 were damaged, but they were on the least vulnerable part of the craft. Even though the exposed part of *Columbia's* hull was glowing red, it was determined that temperatures in those areas would not exceed 600 degrees Fahrenheit, and that was deemed acceptable.

After an eight-hour sleep period, Young and Crippen put *Columbia* through various on-orbit maneuvers, such as turning

The payload bay doors are always open when the shuttle is on orbit so that the heat radiating panels on their interior surfaces can function properly.

the shuttle around and raising and lowering its altitude. They also tested the efficiency of the heat-control system and transmitted television pictures of themselves at work back to earth. They enjoyed pastrami sandwiches for lunch—Young had smuggled them aboard to provide some relief from the monotony of NASA's prepackaged shuttle food—and had a telephone conversation with Vice President George Bush.

On the third and final day of the flight, one of the auxiliary power units (APUs) malfunctioned. These devices control the rudder and flaps, which stabilize the orbiter during reentry. The shuttle has three of them, but if more than one fails, it cannot return to earth safely. Nevertheless, Young initiated de-orbit procedures, instructing the computers to turn *Columbia* around. Flying backward, its rear-mounted orbital maneuvering engines could be fired to slow its speed enough to allow it to descend into the upper atmosphere. The computers then turned the ship around so that it was flying nose forward. At an altitude of four hundred thousand feet, over the east coast of Australia, the atmosphere became dense enough to slow the shuttle further. At that point it became a glider, programmed to descend steeply along a forty-four-hundred-mile path that would bring it down at Edwards Air Force Base in California.

TOUCHDOWN

The temperature of the heat tiles rose steadily until, at 2,750 degrees Fahrenheit, it became hot enough that it ionized the air surrounding it, blanketing the orbiter in an electrically charged cloud. This process, in which the intense heat separates electrons from the atoms that compose the air, was expected, and it caused a sixteen-minute radio blackout with mission control. Communication was reestablished at 180,000 feet. At 130 miles downrange from the landing strip, Young told the computers to carry out a series of S-shaped turns to slow the velocity of the craft to just above the speed of sound.

Thousands of onlookers on the ground heard two loud, sharp sonic booms in rapid succession as *Columbia*, still traveling faster than sound, broke through the sound barrier. Young rolled the orbiter into a looping left turn and aligned it with the runway. The landing gear descended and, as the crowd cheered, the shuttle made a perfect landing. From the cockpit, even Young could not contain his excitement. "This is the world's greatest flying ma-

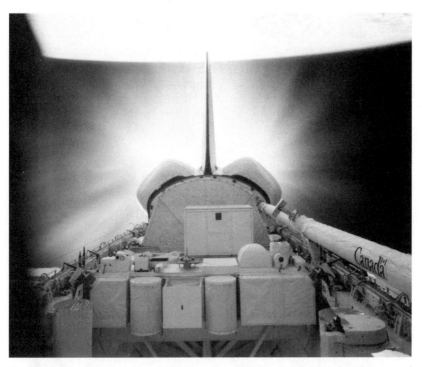

By firing the orbital maneuvering engines (shown here), the shuttle's speed can be slowed enough to reenter the earth's upper atmosphere.

chine!"[72] he shouted over the intercom. The space shuttle's first flight had been a rousing success.

Using a specially designed hoist called a mate-demate device, the orbiter was bolted to the top of a Boeing 747 and flown back to the Kennedy Space Center to be refurbished for its second test flight. During its financial negotiations with Congress and the administration, NASA had promised that it would take no more than two weeks to refit the shuttle and prepare it to launch again after each flight. In fact, it took six months. While that was happening, NASA took careful stock of how the vehicle had performed. Young's and Crippen's major complaint was that the space toilet had failed during the flight, something that sounded trivial but could be a major problem if the shuttle had a full crew of seven and was scheduled for a two-week mission

In addition, analysis revealed that the missing heat tiles had been knocked off by pieces of ice that formed on the liquid fuel tank during launch, a relatively simple problem to fix. Several

THE FIRST SHUTTLE PILOT

Robert Crippen's flight aboard *Columbia* was his first trip into space, although he had been an astronaut in 1969 and had served as part of the ground support team for several earlier NASA missions. He holds an aeronautical engineering degree from the University of Texas, and, like his fellow *Columbia* astronaut John Young, was a navy pilot before joining NASA. As of 1996, he had traveled more than 9.4 million miles in space on four shuttle flights. He was named director of the Kennedy Space Center in 1992, and in 1996 left NASA to become an executive in the aerospace industry.

Robert Crippen's first trip into space was on board Columbia.

minor difficulties with the auxiliary power units and orbital maneuvering engines were also quickly resolved. However, the shock wave produced when the solid rocket boosters ignited was larger than expected, severe enough to damage the delicate scientific instruments the shuttle was expected to routinely carry into orbit. Much of the delay in preparing *Columbia* for its second flight stemmed from attempts to rectify this particular shortcoming.

FLIGHT NUMBER TWO

Columbia's second flight marked the first time in the history of space flight that a vehicle had been reused. The flight was scheduled to last 124 hours and its main purpose was to test the $100 million Canadian-built robotic arm that would enable the shuttle to put satellites into orbit and retrieve them should they need to be repaired or returned to earth. The flight's secondary purpose was to carry a small payload of scientific instruments into space to test the cargo bay's ability to carry delicate payloads safely.

To solve the shock wave problem, NASA engineers installed a huge water tank at the launch pad at a cost of $1.5 million. Just before the solid rocket boosters were set to ignite, this system would release seventy thousand gallons of sound-deadening water per minute into the fire pit at the base of the pad. A test on September 16, 1981, proved that the innovative procedure successfully absorbed the powerful acoustic forces released by the boosters, enabling the shuttle to lift off without damaging its payload.

Several launches had to be aborted because of minor glitches in the onboard computers and the auxiliary power units, but *Columbia* finally blasted off on November 11, with astronauts Joseph Engle and Richard Truly at the controls. Shortly after the shuttle entered orbit, one of the auxiliary power units malfunctioned, forcing mission control to reduce the flight time from 124 to just 54 hours. Engle and Truly, working on an accelerated schedule, managed to complete most of the planned tests and *Columbia* touched down at Edwards Air Force Base on November 14.

Although the flight had been curtailed, *Columbia's* second voyage was instrumental in convincing NASA that the shuttle was approaching operational readiness, the point at which it would be deemed fit to undertake scientific and satellite-launching missions

IF SOMETHING GOES WRONG . . .

Space flight experts, and the astronauts themselves, candidly admit that there is little that can be done to avert disaster should the space shuttle experience the failure of any of its major systems. However, should something go wrong after the solid rocket boosters have been jettisoned, a little more than two minutes following liftoff, the mission can sometimes be aborted safely in one of the following ways:

Return to Launch Site (RTLS): With the main engines still functioning and the external fuel tank still attached, the shuttle would fly downrange, execute a flip maneuver, and fly back toward the Kennedy Space Center. Shortly before landing, it would jettison the external fuel tank into the ocean. This procedure would take about twenty-five minutes.

Transoceanic Abort Landing (TAL): If the problem occurs too late for an RTLS, the shuttle can fly to one of three emergency landing sites located in Spain and Morocco. Shuttle missions are often delayed until weather conditions at these locations allow for the possibility of a safe touchdown. The transatlantic flight would take approximately 35 minutes.

Abort Once Around (AOA): If the shuttle has attained enough velocity to make it into a low orbit before the problem surfaces, it could continue to make one circle around earth and land either at Florida's Kennedy Space Center or at Edwards Air Force Base in California 110 minutes after liftoff.

Abort to Orbit (ATO): If the orbiter is flying fast enough to maintain a low orbit for more than one trip around earth, it could be possible to use its orbital maneuvering system engines to lift it higher and even complete the mission as planned.

for industry and the Department of Defense. The sound absorption system worked as planned and the cargo of scientific instruments experienced no negative effects. There was minimal loss or damage to the heat shield tiles, and the robotic arm indicated it was ready to deploy satellites. Also, preparing *Columbia* for its third flight took ten fewer days than for its second, a small gain in turnaround time that NASA hoped to build on as it gained more experience in refurbishing the complicated vessel. Most important, the agency had proved it was capable of building and operating a reusable space vehicle—even though it had fallen short of the promises it had made in terms of cost and efficiency.

THE FINAL TEST FLIGHTS

The third test flight, launched with only a one-hour delay on March 22, 1982, had astronauts Jack Lousma and Gordon Fullerton on board. The mission was designed to run further tests on the robotic arm and the thermal protection system. The crew also carried out two chemical engineering experiments. On this trip, the shuttle picked up its first space stowaway, a tiny Florida fruit fly, which buzzed around the crew compartment and vanished, never to be seen again.

Except for more damage to the heat tiles, the mission went as planned, but excessively rainy weather at Edwards Air Force Base forced the shuttle to stay in orbit for an extra day and land at the White Sands Missile Range in New Mexico. The change in plans required 550 personnel and 38 train cars of equipment to travel from California to White Sands to prepare the runway and ensure a safe landing.

The fourth and final test flight took place on June 27, 1982. More tests were done on the robotic arm and the shuttle carried the first Department of Defense payload, an infrared telescope that failed to function properly. Apart from that disappointment, for which the Department of Defense accepted the blame, the mission, with astronauts Thomas Mattingly and Henry Hartsfield at the controls, experienced only minor problems. Although the delays and cost overruns were both substantial, the shuttle was finally deemed operational.

Author Richard Lewis comments:

> In all major respects, the shuttle was working as it was designed to do. The great unknowns—the performance of a winged spacecraft during entry into the atmosphere, the validity of the ceramic heat shield, and the aerospace craft's handling in the atmosphere as an unpowered glider—had been resolved.[73]

THE SHUTTLE
COMES OF AGE

Between 1982 and 1986 the space shuttle flew twenty-four successful missions. Each flight encountered problems with various
aspects of the complex system, but each was resolved—at least
well enough to enable the following flight to take place. Three
new orbiters—*Challenger*, *Discovery*, and *Atlantis*—completed
the fleet. But the promises NASA had made to keep the program funded began to create new problems. The agency had
told Congress and the federal administration that the shuttle
would cost $10.4 million per flight, which would be paid by De-
partment of Defense contracts and fees for launching commercial satellites for private enterprise. In effect, NASA had said,
once the shuttle was built it would pay for itself. By the time the
craft became operational in 1982, however, it was clear that the
actual cost per flight would be around $155 million because of
the time and manpower required to refurbish each of the orbiters between missions. The most that NASA could charge its
paying customers was approximately $70 million per launch;
otherwise the agency would lose business to other highly competitive unmanned satellite-launching rockets.

Technical problems also began to mount, especially with the
main engines, which had been designed to make fifty-five
flights before requiring a major overhaul. As it turned out, the
engines had to be extensively rebuilt after every three flights.
Because they cost $36 million each, budget restraints meant
there were few spares; the launch schedule quickly began to fall
behind.

Other aspects of the system also showed the effects of earlier
cost cutting. In 1983 *Challenger* was almost lost when the lining
of the exhaust nozzle on one of its solid rocket boosters eroded.
Had it been required to burn for eight seconds longer than it
did, the booster would have experienced a catastrophic explosion. The shuttle's reliability was also criticized, particularly by

the Department of Defense. Secretary Caspar Weinberger of the Department of Defense endorsed a report stating that exclusive reliance on the shuttle for the department's space needs had come to represent "an unacceptable national security risk."[74] Yet NASA administrator James Beggs retorted that the shuttle "is the most reliable space transportation system ever built."[75] Weinberger was unmoved, and announced that the Defense Department would use the unmanned Titan III rocket for many of its key satellite launches.

Nevertheless, NASA continued to make minor modifications to the shuttle (to correct problems as they were identified) and fought hard to win a larger share of the commercial satellite-launching business. Utah senator Jake Garn and Representative Bill Nelson of Florida, both strong supporters of the space agency, flew on shuttle missions, convinced that the vessel was safe enough to carry nonprofessional astronauts. Finally, in 1985, for the first time, more shuttles than expendable rockets were launched. "The shuttle," says T. A. Heppenheimer, "was indeed approaching routine operation."[76]

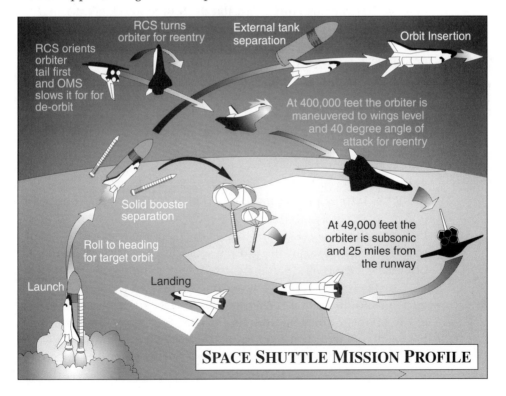

SPACE SHUTTLE MISSION PROFILE

THE CHALLENGER DISASTER

Unsteady as it had been, the progress of the shuttle toward the goal of making space flight a routine activity came to an explosive end on January 28, 1986. That day, on the twenty-fifth shuttle flight, the *Challenger*, with a crew that included schoolteacher Christa McAuliffe, erupted into an orange fireball seventy-three seconds after liftoff. The crew compartment fell nine miles into the Atlantic Ocean, killing everyone on board.

The disaster was caused by a malfunction of the seals that helped hold the three segments of the solid rocket boosters together. Each of the seals was surrounded by two thick rubber O-rings designed to prevent gases from the burning solid rocket fuel from escaping and damaging the rest of the shuttle. NASA was aware as early as the second test flight that the O-rings had significant design shortcomings, but for budgetary reasons—and to adhere to a strict launch schedule—the agency had failed to correct them. One researcher admits:

> O-ring anomalies had been detected to varying degrees on 12 previous flights. Erosion had been detected on all 12 occasions, and gas actually escaped on 10 of them. All of these anomalies had been recorded upon occurrence, and this data was known at the Flight Readiness Review for STS 33/51-L [the *Challenger* launch].[77]

THE VEHICLE ASSEMBLY BUILDING AND THE CRAWLER TRANSPORT

Prior to takeoff the shuttle is assembled in a large structure called the Vehicle Assembly Building (VAB). Behind the VAB's massive 456-foot doors, the largest doors in the world, the orbiter, external tank, and solid rocket boosters are hoisted by 250-ton cranes and bolted together. The shuttle is then placed on a mobile launcher platform. The total assembly, weighing 12 million pounds, is carried to the launch pad by a crawler transport at a painstakingly slow speed of one mile per hour. The trip takes about five hours, during which the crawler consumes diesel fuel at a rate of 150 gallons per mile.

The problems had happened primarily when the shuttle launched during lower than normal temperatures, and Thiokol, the manufacturer of the solid rocket boosters, had made NASA aware of this potential for disaster. In July 1985 Roger Boisjoly, Thiokol's senior O-ring specialist, wrote his concerns about the O-ring problem in a memo to his superiors. "It is my honest and very real fear that if we do not take immediate action to dedicate a team to solve the problem, we stand in jeopardy of losing a flight along with all the launch pad facilities."[78]

NASA and Thiokol's upper-level management responded with nonchalance. "There was no real urgency," says T. A. Heppenheimer. "Better seals were just one more improvement that would come along in good time, as the shuttle design continued to improve."[79]

As *Challenger*'s launch date approached, Florida experienced an unusual cold snap. Both Thiokol and the NASA executive in charge of Thiokol's contract wanted to push ahead over the objections of the company's engineers, and concerns about the O-rings' safety were not passed along to the team at the Kennedy Space Center who would make the final decision to launch or not to launch. "The experience of two dozen successful Shuttle flights had bred complacency," Heppenheimer says. "The seals had worked before, however imperfectly; why not again?"[80]

THE ROGERS COMMISSION

On February 3, 1986, President Ronald Reagan called a halt to all shuttle flights for at least two years and appointed a commission to investigate the *Challenger* accident. The panel—the Rogers Commission—was headed by former secretary of state William Rogers and included, among other scientists and aviation experts, astronauts Neil Armstrong and Sally Ride, Brigadier General Charles Yeager—the first man to break the sound barrier—and Nobel Prize–winning physicist Richard Feynman. Their concluding report stated, in part:

> The consensus of the Commission and participating investigative agencies is that the loss of the Space Shuttle *Challenger* was caused by a failure in the joint between the two lower segments of the right Solid Rocket Motor. The specific failure was the destruction of the seals that are intended to prevent hot gases from leaking through

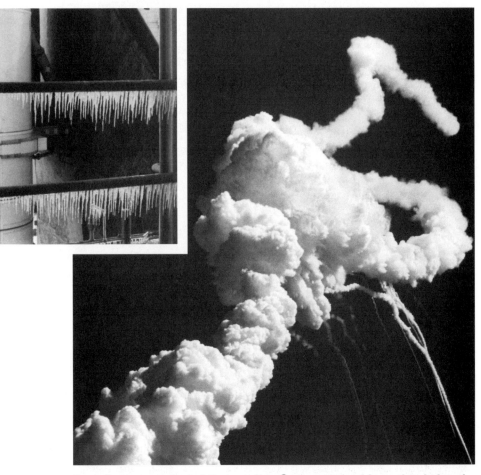

Investigators concluded that eroding O-rings and cold weather led to the 1986 Challenger *explosion.*

the joint during the propellant burn of the rocket motor. The evidence assembled by the Commission indicates that no other element of the Space Shuttle system contributed to this failure.[81]

The report went on to criticize both NASA and Thiokol for ignoring danger signs that the O-rings were headed for a catastrophic failure. As the problems "grew in number and severity," the commissioners wrote, "NASA minimized them in briefings and reports; as tests and then flights confirmed damage to the sealing rings, the reaction by both NASA and Thiokol was to increase the amount of damage considered 'acceptable.'"[82]

In an addendum to the report, Richard Feynman flatly accused NASA of jeopardizing the lives of astronauts by ignoring the advice of its own engineers, who on a number of occasions had estimated that under the pressure of budgetary limitations and an overambitious launch schedule, one shuttle flight in every one hundred would end in disaster. Feynman wrote:

> Official management, on the other hand, claims to believe the probability of failure is a thousand times less. One reason for this may be an attempt to assure the government of NASA perfection and success in order to ensure the supply of funds. . . . In any event this has had very unfortunate consequences, the most serious of which is to encourage ordinary citizens to fly in such a dangerous machine, as if it had attained the safety of an ordinary airliner . . . [NASA] must live in reality in comparing the costs and utility of the Shuttle to other methods of entering space. And they [NASA] must be realistic in making contracts, in estimating costs, and the difficulty of the projects. . . . NASA owes it to the citizens from whom it asks for support to be frank, honest, and

Richard Feynman (background) accused NASA of endangering the lives of astronauts by disregarding the agency's own engineers' advice.

informative, so that these citizens can make the wisest decisions for the use of their limited resources. For a successful technology, reality must take precedence over public relations, for nature cannot be fooled.[83]

THE REDESIGN OF THE SHUTTLE

Following the *Challenger* explosion, the shuttle did not fly again until September 1988. During the time when shuttle flights were grounded, the craft was extensively redesigned based on recommendations made by the Rogers Commission, whose members found serious flaws in almost every aspect of the system. The solid rocket boosters underwent changes in seven key areas apart from the O-rings; these included improvements to the ignition system, the nozzle through which the propellant was expelled to provide lifting power, and points at which the booster segments were joined together. New, more stringent testing procedures were also established. The O-rings themselves were equipped with electric heaters to enable them to function safely through a wider range of temperatures.

The main engines were also found to have weaknesses that limited their durability. Sixteen previous near disastrous failures had occurred, and crucial components routinely cracked under stress. NASA was forced to recognize that the engines would require more maintenance than the agency had previously admitted, and procedures were implemented to improve reliability without compromising performance.

The orbiter underwent seventy-six modifications, including a new escape system for the crew, augmented thermal protection on the underside of the forward fuselage, new brakes to ensure safer landings, and a more reliable mechanism for detaching the external fuel tank in flight. The escape system now allows astronauts to bail out of the orbiter under low-speed conditions, giving them a better chance of avoiding the massive impact that killed the *Challenger* crew. Changes to the computer software allows the orbiter to fly with more stability, maximizing the amount of time for bailout. An inflatable escape slide for use during ground evacuations was also added.

Further, the commission insisted that NASA extensively revamp its management structure to ensure better communication between engineers and high-level executives. As a result, an

THE RUSSIAN SHUTTLE

In response to the American shuttle program, the Russian (then Soviet) government undertook the task of building its own version—the *Buran* (Russian for snowstorm) in 1976. They built two Burans, which closely resembled the U.S. shuttle, one for nonorbital test flights and the other for full operational use. The *Buran* flew only one orbital mission, controlled from the ground without an onboard crew, in 1988 before the Soviet space program ran out of money for further tests. The nonorbital version was sold to an amusement company that converted it into a tourist restaurant in Moscow's Gorky Park. The fully operational model has been put in storage, and the factory where they were both built now manufactures buses, hypodermic syringes, and diapers.

Russia's space shuttle Buran *flew only one orbital mission.*

independent flight-readiness panel now reviews all launch decisions, and flights are scheduled on a more realistic basis. But the most far-reaching effect of the *Challenger* accident is that the shuttle was forbidden from lifting commercial satellites into orbit and banned from carrying passengers who were not fully trained professional NASA astronauts. These two limitations dashed forever the agency's dream that the shuttle would be able to earn enough money from private industry to pay for itself. Instead, it became strictly a government-funded, government-sponsored project.

THE FUTURE

Since the shuttle resumed flying in 1988, it has carried out a number of Defense Department and scientific missions, including putting the Hubble Space Telescope into orbit and repairing

Astronauts release the Hubble Space Telescope from above the cargo bay of the space shuttle.

it after it became apparent that a badly designed lens was critically limiting its usefulness. The shuttle was the only vehicle in the history of space travel versatile enough to have carried out the delicate work required to fix the Hubble, and in doing so it saved American taxpayers billions of dollars. But, even though restrictions on commercial satellite launches were eased, the space shuttle has effectively been eclipsed by cheaper and safer unmanned rockets.

Two developments during the 1990s have given the shuttle an expanded role during space exploration. Beginning in the mid-1990s, NASA commissioned research into the X-33, a rocket plane that would be part of a fully reusable space transportation system reminiscent of the early designs for the shuttle. Once again, budget problems emerged, and in 1999 the agency shelved all plans for the project. That decision guaranteed that the shuttle would be America's only manned space vehicle in use until at least the year 2020.

The second event to have a major impact on the shuttle's future was the decision to build the International Space Station. In 1993, after heavy lobbying by President Bill Clinton, Congress narrowly approved a bill allowing the United States to enter into a partnership with Russia and other nations to proceed with the construction of the station. The shuttle played a major role in transporting the components of the station into space, assembling them while on orbit, and supplying the station once it became operational. After three turbulent decades, the shuttle finally had the job it was originally conceived to perform.

A new orbiter, the *Endeavour,* was added to the fleet in 1991 to replace the *Challenger.* And as of May 2001, NASA was in the process of enhancing the safety and performance of the entire fleet of shuttles to meet its twenty-first-century obligations. Planned projects included equipping the main engines with

In 1991, NASA officials replaced the Challenger *with a new shuttle, the* Endeavour.

CAN THE SHUTTLE BE SEEN FROM EARTH?

Under certain conditions the shuttle is often visible from earth, appearing like a bright point of light moving quickly across the sky. During reentry, it becomes a dramatically blazing fireball, clearly visible to all in the vicinity of its flight path. NASA's website provides information on the best viewing times, but as a rule the shuttle is visible only for a few minutes at a time and only on cloudless nights. The shuttle moves very fast (it orbits earth once every ninety minutes), but using a pair of good binoculars, observers have been able to see details of the orbiter's shape and structure. Occasionally, the orbiter eliminates waste water by spraying it into space, where it immediately freezes into highly visible ice crystals. Another good opportunity to view the shuttle is when it docks with the International Space Station. The size of the combined vehicles boosts their visibility.

what a NASA fact sheet calls "a high-tech optical and vibration sensor system and computing power in the engines that will 'see' trouble coming a fraction of a second before it can do harm."[84] Also, the engine's main combustion chamber will be enlarged to reduce stress and a simplified nozzle will eliminate more than five hundred places where the metal components have to be welded together, each of which is a potential leaking point. Further, electrical generators for the hydraulic system will make it no longer necessary for the orbiter to carry volatile rocket fuel for on-orbit maneuvers, and computer enhancements will reduce the pilot's workload during an emergency, allowing the crew to concentrate on implementing escape procedures. The solid rocket boosters are also being redesigned to increase their ability to withstand the pressure generated by the combustion of fuel. The improvements, NASA says, will "double the shuttle's safety by 2005."[85]

IS THE SHUTTLE SAFE?

The shuttle in 2001 is far safer than it was before the *Challenger* explosion in 1986. But experts agree that NASA's promise that the shuttle would make space travel safe and routine has not been kept. Space travel, they say, is inherently risky and always will be. Danger can be minimized, but not eliminated.

Bill Nelson was a congressman from Florida when he flew on the shuttle mission launched immediately prior to the *Challenger*. Several years following the disaster, he reached back into Greek mythology to explain the shuttle's development.

> The Space Shuttle represents a revolutionary departure from the expendable rockets that were the launchers for the first manned flights. But the triumph has not been without tragedy. Such tragedy in the heavens was foretold as far back as Greek mythology. On the island of Crete in ancient Greece, as legend has it, two of King Minos's subjects [Daedalus and his son Icarus] wanted to fly. They constructed wings made of wax and feathers. Daedalus and Icarus soared about the sky, defying the ancient mythical gods. Icarus . . . wanted to go higher and higher. Daedalus warned him against this, but the young man's adventurous spirit overrode his obedience to his father. As he flew higher, he passed too close to the sun. The wax holding the feathers in his wings melted, and he fell into the sea. Man [like Icarus] will continue to go higher and higher, but not without sometimes falling back into the sea—as *Challenger* did. Tragedy is always a companion of triumph and is the price that must be paid by all who venture beyond the known into the unknown.[86]

Notes

Introduction

1. Richard Lewis, *The Voyages of Columbia*. New York: Columbia University Press, 1984, pp. vii–viii.
2. Quoted in Dan Falk, "Shuttle," CBC News, 2001. www.cbc.ca/news/indepth/shuttle/.
3. Lewis, *The Voyages of Columbia*, p. viii.

Chapter 1: The Birth of the Shuttle

4. John F. Kennedy, Address at Rice University on the Space Effort, September 12, 1962, Rice University Archives. www.riceinfo.rice.edu.
5. Project Apollo: A Retrospective Analysis. www.hq.nasa.gov/office/pao/history/apollomom/Apollo.html
6. Lewis, *The Voyages of Columbia*, p. 17.
7. The National Aeronautics and Space Administration www.hq.nasa.gov/pao/history.
8. David M. Harland, *The Space Shuttle: Roles, Missions and Accomplishments*. Chichester, England: Praxis Publishing, 1998, p. xiii.
9. Quoted in Dennis R. Jenkins, *Space Shuttle: The History of Developing the National Space Transportation System*. Melbourne, FL: Dennis R. Jenkins, 2000, p. 3.
10. Harland, *The Space Shuttle*, p. xiii.
11. Harland, *The Space Shuttle*, p. 3.
12. Lewis, *The Voyages of Columbia*, p. 19.
13. Joan Lisa Bromberg, *NASA and the Space Industry*. Baltimore: Johns Hopkins University Press, 1999, p. 77.
14. Report of the Space Task Group, 1969. www.hq.nasa.gov/office/pao/history/taskgrp.html.
15. Andrew J. Dunbar and Stephen P. Waring, *The Power to Explore: A History of the Marshall Space Flight Center*. NASA, 1991. www.history.msfc.nasa.gov/book/bookcover.html.
16. Quoted in Dunbar and Waring, *The Power to Explore*, p. 276.
17. Lewis, *The Voyages of Columbia*, p. 24.
18. Lewis, *The Voyages of Columbia*, p. 24.
19. Lewis, *The Voyages of Columbia*, p. 25.

20. Quoted in Lewis, *The Voyages of Columbia*, p. 25.

Chapter 2: The Battle to Save the Shuttle

21. T. A. Heppenheimer, *Countdown: A History of Space Flight*. New York: John Wiley & Sons, 1997, p. 250.

22. OMB's Role, Office of Management and Budget. www. whitehouse.gov/OMB/organization/role.html.

23. Heppenheimer, *Countdown*, p. 250.

24. Howard E. McCurdy, *Inside NASA: High Technology and Organizational Change in the U.S. Space Program*. Baltimore: Johns Hopkins University Press, 1993, p. 85.

25. Lewis, *The Voyages of Columbia*, p. 25.

26. Lewis, *The Voyages of Columbia*, p. 25.

27. Quoted in Lewis, *The Voyages of Columbia*, p. 26.

28. Bromberg, *NASA and the Space Industry*, p. 75.

29. Bromberg, *NASA and the Space Industry*, pp. 75–76.

30. Quoted in Heppenheimer, *Countdown*, p. 253.

31. Lewis, *The Voyages of Columbia*, p. 28.

32. Heppenheimer, *Countdown*, p. 250.

33. Heppenheimer, *Countdown*, p. 253.

34. Quoted in Jenkins, *Space Shuttle*, p. 76.

35. Heppenheimer, *Countdown*, p. 255.

36. Bromberg, *NASA and the Space Industry*, p. 87.

37. Quoted in Heppenheimer, *Countdown*, p. 256.

38. Quoted in McCurdy, *Inside NASA*, p. 86.

Chapter 3: The Shuttle Takes Shape

39. David Baker, *Space Shuttle*. New York: Crown Publishers, 1979, p. 6.

40. Tom D. Crouch, *Aiming for the Stars: The Dreamers and Doers of the Space Age*, Washington, DC: Smithsonian Institution Press, 1999, p. 251.

41. Heppenheimer, *Countdown*, p. 259.

42. Cliff Lethbridge, Space Shuttle Program Background, 1998. www.spaceline.org.

43. Heppenheimer, *Countdown*, p. 260.

44. Wernher von Braun, *Space Travel: A History*. Ed. David Dooling. New York: Harper & Row, 1985, p. 242.

45. Bromberg, *NASA and the Space Industry*, pp. 90, 93.

46. Bromberg, *NASA and the Space Industry*, p. 95.

47. Bromberg, *NASA and the Space Industry*, p. 99.

48. Bromberg, *NASA and the Space Industry*, p. 99.

49. Bromberg, *NASA and the Space Industry*, p. 100.

50. NASA Space Shuttle: Orbiter Structure. www.spaceflight. nasa.gov/shuttle/reference.

51. Bromberg, *NASA and the Space Industry*, pp. 100–101.

Chapter 4: Lessons in Reality

52. Lewis, *The Voyages of Columbia*, p. 34.

53. Lewis, *The Voyages of Columbia*, p. 33.

54. Quoted in Lewis, *The Voyages of Columbia*, p. 60.

55. Lewis, *The Voyages of Columbia*, p. 62.

56. Lewis, *The Voyages of Columbia*, p. 63.

57. Lewis, *The Voyages of Columbia*, p. 64.

58. Heppenheimer, *Countdown*, pp. 308-309.

59. Heppenheimer, *Countdown*, p. 309.

60. Crouch, *Aiming for the Stars*, p. 255.

61. Jenkins, *Space Shuttle*, p. 169.

62. Jenkins, *Space Shuttle*, p. 168.

63. Lewis, *The Voyages of Columbia*, pp. 36–37.

64. Lewis, *The Voyages of Columbia*, p. 88.

65. Lewis, *The Voyages of Columbia*, p. 89.

66. Lewis, *The Voyages of Columbia*, p. 88.

67. Lewis, *The Voyages of Columbia*, p. 91.

Chapter 5: The Shuttle Blasts Off—Finally

68. Quoted in Lewis, *The Voyages of Columbia*, p. 155.

69. Lewis, *The Voyages of Columbia*, p. 155.

70. Quoted in Lewis, *The Voyages of Columbia*, p. 128.

71. Quoted in Lewis, *The Voyages of Columbia*, p. 130.

72. Quoted in Lewis, *The Voyages of Columbia*, p. 143.

73. Lewis, *The Voyages of Columbia*, p. 181.

Chapter 6: The Shuttle Comes of Age

74. Quoted in Heppenheimer, *Countdown*, p. 322.

75. Quoted in Heppenheimer, *Countdown*, p. 322.

76. Heppenheimer, *Countdown*, p. 324.

77. Jenkins, *Space Shuttle*, p. 279.

78. Quoted in Heppenheimer, *Countdown*, p. 324.

79. Heppenheimer, *Countdown*, pp. 324–325.

80. Heppenheimer, *Countdown*, p. 325.

81. Quoted in Jenkins, *Space Shuttle*, p. 279.

82. Quoted in Heppenheimer, *Countdown*, p. 326.

83. Richard Feynman, *The Pleasure of Finding Things Out*. Ed. Jeffrey Robbins. Cambridge, MA: Helix Books, 1999, pp. 168–169.

84. 21st Century Space Shuttle, NASA, 2000, p. 1. www. spaceflight.nasa.gov/shuttle/upgrade.

85. 21st Century Space Shuttle, NASA. www.spaceflight.nasa. gov/shuttle/upgrade.

86. Bill Nelson with Jamie Buckingham, *Mission: An American Congressman's Voyage to Space*. Orlando, FL: Harcourt Brace Jovanovich, 1988, p. 291.

For Further Reading

Wendy Baker, *America in Space*. New York: Crescent Books, 1986. Concise, well-illustrated overview of the American space program from its beginning to the *Challenger* explosion in 1986.

Robert Gardner, *Space*. New York: Twenty-First Century Books, 1994. Straightforward explanation of the science behind space flight in general and the shuttle in particular.

Wayne Lee, *To Rise from Earth: An Easy to Understand Guide to Space Flight*. New York: Facts On File, 1995. Written by an engineer at NASA's Jet Propulsion Laboratory, this book presents the technical aspects of the shuttle's development and operation in layman's language.

Richard Lewis, *The Voyages of Columbia*. New York: Columbia University Press, 1984. Lewis, a longtime space journalist, tells the story of the birth of the shuttle in nontechnical language. He also has a newsperson's instinct for dramatic detail and revealing anecdotes.

⌥ Works Consulted

Books

David Baker, *Space Shuttle*. New York: Crown Publishers, 1979. This account of the development of the shuttle is dated, but it provides a good description of the various components that make up the system and the technical landscape that existed when the shuttle was in the design stage.

Joan Lisa Bromberg, *NASA and the Space Industry*. Baltimore: Johns Hopkins University Press, 1999. Bromberg analyzes the complex relationship between NASA and the aerospace industry, providing background for some of the decisions made in building the shuttle.

Tom D. Crouch, *Aiming for the Stars: The Dreamers and Doers of the Space Age*, Washington, DC: Smithsonian Institution Press, 1999. A nontechnical history of the U.S. space program that puts the shuttle in context.

Andrew J. Dunbar and Stephen P. Waring, *The Power to Explore: A History of the Marshall Space Flight Center*. NASA, 1991. The Marshall Space Flight Center played a key role in the development of the shuttle, especially the design and testing of the main engines. The story of the problems encountered is told briefly in this history of the center.

Richard Feynman, *The Pleasure of Finding Things Out*. Ed. Jeffrey Robbins. Cambridge, MA: Helix Books, 1999. Feynman, a Nobel Prize–winning physicist is highly critical of NASA's role in the *Challenger* accident. He focuses on shortcomings in the design and testing of the main engines and solid rocket boosters.

David M. Harland, *The Space Shuttle: Roles, Missions and Accomplishments*. Chichester, England: Praxis Publishing, 1998. Harland focuses more on the shuttle's missions than on its development, but the first two chapters provide a brief overview of the relevant material.

T. A. Heppenheimer, *Countdown: A History of Space Flight*. New York: John Wiley & Sons, 1997. Heppenheimer deals more with the political and economic problems that had to be overcome in building the shuttle than he does with the technical difficulties.

Dennis R. Jenkins, *Space Shuttle: The History of Developing the National Space Transportation System*. Melbourne, FL: Dennis R. Jenkins, 2000. A detailed and highly technical account of the problems encountered in designing and building the shuttle. Perhaps the most thorough coverage of the material currently in print.

Howard E. McCurdy, *Inside NASA: High Technology and Organizational Change in the U.S. Space Program*. Baltimore: Johns Hopkins University Press, 1993. Critical account of communications and decision making within the space agency, in part as these phenomena affected building the shuttle.

NASA, *21st Century Space Shuttle*. NASA, 2000. A brief, well-illustrated overview of shuttle improvements on the drawing board in the latter half of the year 2000. All the proposals are subject to budget approval.

Bill Nelson with Jamie Buckingham, *Mission: An American Congressman's Voyage to Space*. Orlando, FL: Harcourt Brace Jovanovich, 1988. Nelson flew on the shuttle immediately preceding the ill-fated *Challenger* flight and provides a balanced assessment of shortcomings in the shuttle program from a politician's perspective.

Wernher von Braun, *Space Travel: A History*. Ed. David Dooling. New York: Harper & Row, 1985. Von Braun presents an opinionated insider's view of the development of the shuttle and other spacecraft. Much of the material on the shuttle was compiled from his notes after his death.

Internet Sources

Report of the Space Task Group, 1969, www.hq.nasa.gov/office/pao/history/taskgrp.html.

Dan Falk, "Shuttle," CBC News, 2001 www.cbc.ca/news/indepth/shuttle.

Cliff Lethridge, Space Shuttle Program Background, 1998. www.spaceline.org.

John F. Kennedy, Address at Rice University on the Space Effort, September 12, 1962, Rice University Archives. www.riceinfo.rice.edu.

NASA Space Shuttle: Orbiter Structure. www.spaceflight.nasa.gov/shuttle/reference.

OMB's Role, Office of Management and Budget. www. whitehouse.gov/OMB/organization/role.html.

Project Apollo: A Retrospective Analysis. www.hq.nasa.gov/ office/pao/history/apollomom/Apollo.html.

www.nasa.gov. NASA operates a large number of websites dealing with its many programs and centers, including the space shuttle. These sites contain much information, but they are not well organized, so it takes some effort to find specific pieces of information. The various search engines are somewhat helpful.

INDEX

redesign of, 92
manned space vehicles, 30
Marshall Space Flight Center, 37, 47
Martin Marietta, 49–50, 64
Mathmatica Inc., 43
Mattingly, Thomas, 85
McAuliffe, Christa, 88
McCurdy, Howard E., 28, 40
McDonnell-Douglas Astronautics Company, 41, 48
middeck, 51–52
mission abortion, 84
Mondale, Walter, 31–32, 33, 39, 58
moon landing, 14
motion, law of, 34
Mueller, George, 23–24, 34
Muskie, Edmund, 58
Myers, Dale, 58

NASA
    alliance between air force and, 34–36, 37, 38
    claims made about space shuttle by, 11, 55, 57
    compromises made by, 32–33, 37–38, 40
    criticism of, after *Challenger* disaster, 90–92
    funding obstacles faced by, 12, 23, 26–30, 32–34, 42–49, 61, 70–71
    fund-raising campaign of, 20–21
    growth of, during Apollo, 16–17
    plans for space station, 20–21
    political opposition to, 12, 28, 31–34, 39, 58
    revamping of, after *Challenger* disaster, 92–93
National Aeronautics and Space Administration

(NASA). *See* NASA
navigation system, 66–67
Nelson, Bill, 87, 97
Newton, Isaac, 34
Nixon, Richard
    opposition to space shuttle by, 31, 32
    support for space shuttle by, 8, 20–21, 39, 44–46
North American Rockwell Corporation, 41, 48

Office of Management and Budget (OMB), 26–27, 42, 43
OMB. *See* Office of Management and Budget
orbiter
    contract for, 48–49
    design of, 12, 24–25, 37, 41, 51–53, 65–66
    redesign of, 92
O-rings, 88–89, 90, 92

Page, George, 76
Paine, Thomas
    air force support sought by, 34–36
    compromises made by, 32
    funding obstacles faced by, 32–34
    resignation of, 43
    space station proposal of, 20–21
payload bay, 52
Perry, William, 71
political considerations, 28, 31–34, 39, 58
Pratt & Whitney, 48

Reagan, Ronald, 73, 89
Remote Manipulator System (RMS), 52
Return to Launch Site (RTLS), 84

PICTURE CREDITS

Cover photos: NASA (all)
© AFP/CORBIS, 65, 79
© James L. Amos/CORBIS, 56
© Bettmann/CORBIS, 18 (left), 21, 23, 49, 74, 81, 91
© CORBIS, 45, 71, 77, 90 (right)
Express Newspapers/G832/Archive Photos, 15
© Owen Franken/CORBIS, 32
Bernard Gotfryd/Archive Photos, 63
© George Hall/CORBIS, 42 (top)
Hulton/Archive by Getty Images, 8, 9
© Hulton-Deutsch Collection/CORBIS, 82
Chris Jouan, 52, 54, 87
© Museum of Flight/CORBIS, 42 (bottom)
NASA, 11, 13, 16, 27, 29, 33, 60, 67, 90 (left)
© NASA/Roger Ressmeyer/CORBIS, 18 (right), 50, 94
© Roger Ressmeyer/CORBIS, 38, 59, 76, 93
Reuters/Duffin McGee/Archive Photos, 35
Reuters/Joe Skipper/Archive Photos, 44
USSEA/Space Age Times, 95

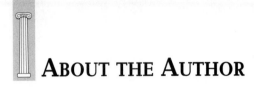

ABOUT THE AUTHOR

Robert Taylor has written on science, technology, history, politics, law, philosophy, medicine and contemporary culture. He lives in West Palm Beach, Florida.

LAKE OSWEGO JR. HIGH SCHOOL
2500 S.W. COUNTRY CLUB RD.
LAKE OSWEGO, OR 97034
635-0335